Cell Migration: Signalling and Mechanisms

Translational Research in Biomedicine

Vol. 2

Series Editor

Samuel H.H. Chan Kaohsiung

Associate Editor

Julie Y.H. Chan Kaohsiung

Cell Migration: Signalling and Mechanisms

Volume Editors

Frank Entschladen Witten
Kurt S. Zänker Witten

20 figures and 2 tables, 2010

KARGER Basel · Freiburg · Paris · London · New York · Bangalore · Bangkok · Shanghai · Singapore · Tokyo · Sydney

Prof. Dr. Frank Entschladen
Institute of Immunology
University of Witten/Herdecke
DE–58448 Witten (Germany)

Prof. Dr. Kurt S. Zänker
Institute of Immunology
University of Witten/Herdecke
DE–58448 Witten (Germany)

Library of Congress Cataloging-in-Publication Data

Cell migration : signalling and mechanisms / volume editors, Frank Entschladen, Kurt S. Zänker.
p. ; cm. -- (Translational research in biomedicine, ISSN 1662-405X ; v. 2)
Includes bibliographical references and indexes.
ISBN 978-3-8055-9321-2 (hard cover : alk. paper)
1. Cells migration. 2. Cellular signal transduction. I. Entschladen, Frank. II. Zänker, Kurt S. III. Series: Translational research in biomedicine, v. 2. 1662-405X ;
[DNLM: 1. Cell Movement. 2. Signal Transduction--physiology. QU 375 C3931 2010]
QH647.C439 2010
571.6'7--dc22

2009043614

www.karger.com
Printed in Switzerland on acid-free and non-aging paper (ISO 9706) by Reinhardt Druck, Basel
ISSN 1662–405X
ISBN 978–3–8055–9321–2
e-ISBN 978–3–8055–9322–9

Contents

Foreword

Welcome to volume two of *Translational Research in Biomedicine*, a monograph series dedicated to the dissemination of seminal information in contemporary biomedicine with a translational orientation.

As I pointed out in the inaugurating volume, translational research (TR) is now a household word in the arena of contemporary biomedical research, although a universal definition for this term is currently wanting. In a more restricted sense, TR is often associated with research and development based on the classical bench to bedside approach. Thus, it has been said that 'the goal of TR is to implement in vivo measurements and leverage preclinical models that more accurately predict drug effects in humans' [1]; or 'TR describes a unidirectional effort to test in humans novel therapeutic strategies developed through experimentation' [2]. The current enthusiasm over the application of genomic or stem cell research to therapeutic strategies is also grounded on a similar premise. In a broader sense, TR is taken as a bench to bedside and back approach to foster communication between the scientific community and clinical practitioners [1]. It is a concept that needs the attention from everyone and should be the foundation of a modern understanding of health provision [3].

If we subscribe to the philosophical connotation that medical research is for the betterment of humankind, then we should realize that there is no real demarcation between clinical (bedside) and preclinical (bench) research. This is because the only difference is that human subjects instead of animals, tissues or cells are employed in the studies. Nonetheless, governed by the same ethical principles and guidelines, all of them will reveal information in some aspects of biomedicine. Thus, this monograph series shall take a holistic view on TR that transcends the boundaries between bench and bedside research. Each volume shall be a synthesis of ideas, technologies and research outcomes that are associated with a particular theme in contemporary biomedicine, to be edited by experts in that field. The word 'translation' is most commonly defined as expression of words in another language. Its definitions can be extended to encompass expression in simpler language and uncomplicated interpretation. In this spirit, all chapters in this series will be presented in a fashion that is amenable to non-experts, be they scientists or clinical practitioners.

My sincere thanks go to Professors Frank Entschladen and Kurt S. Zänker, whose patience and expert efforts have made this timely volume on 'Cell Migration: Signalling and Mechanisms' a reality. I also wish to acknowledge the capable hands of Stefan Goldbach and Ruedi Jappert at S. Karger AG during the development and production of this series. Last but not least, the publication of *Translational Research in Biomedicine* would not have been possible without the foresight, enthusiasm and whole-hearted support of my dear friend, Dr. Thomas Karger.

Samuel H.H. Chan, Kaohsiung
Series Editor

References

1 Hörig H, Pullman W: From bench to clinic and back: perspective on the First IQPC Translational Research Conference. J Transl Med 2004;2:44–51.
2 Mankoff SP, Brander B, Ferrone S, Marincola FM: Lost in translation: obstacles to translational medicine. J Transl Med 2004;2:14–18.
3 Sonntag KC Implementations of translational medicine. J Transl Med 2005;3:33–35.

Preface

Towards the Future of Cell Migration Research

As often stated, cellular migration is the crown achievement in biology. The chapters of this book highlight cell behavior with respect to inducing, controlling and terminating the mechanisms of cell migration. This book was only made possible because leading experts from different relevant disciplines contributed to the state of the art of this most fascinating field at the frontier of biological network research with special reference to the field of stem cells, tumor cells and immune competent cells.

Cellular migration research, which has in part been pioneered by the continuous contributions of the Institute of Immunology and Experimental Oncology, University Witten/Herdecke over the past 20 years, is now an area which is rapidly expanding and attracting an increased amount of interest among a broad audience of scientists and clinicians. Besides other cellular features such as proliferation, differentiation and apoptosis, the key elements of cellular migration are now focused on for their potentials in health and disease. Here, we provide a wide and updated view on the major mechanisms involved in cell migration.

We have been fortunate to recruit eminent scientists from around the world who give overviews in their fields of expertise. We would like to thank these distinguished authors for their contributions, which we hope will give the readers a sufficient and fascinating insight into this biological issue – cell migration.

We are grateful to Karger Publishers and to Samuel H.H. Chan, the Series Editor of the newly established and important series *Translational Research in Biomedicine* for publishing this volume. It is our belief that, at the beginning of this new series of publications, this added volume will provide interesting and thought-provoking aspects on this fundamental part of cell behavior in health and disease.

Frank Entschladen and *Kurt S. Zänker*, Witten
Volume Editors

Entschladen F, Zänker KS (eds): Cell Migration: Signalling and Mechanisms.
Transl Res Biomed. Basel, Karger, 2010, vol 2, pp 1–6

The Migrating Cell

Frank Entschladen · Kurt S. Zänker

Institute of Immunology, Witten/Herdecke University, Witten, Germany

Abstract

Cell migration is a complex coordinated process in which several compartments of the cells are involved, including surface receptors, signaling elements, and the cytoskeleton. Collectively, these interacting components can be termed as the cell's migrosome. Although the principal mechanisms of the migration are the same among the cells that migrate in the adult human body, i.e. leukocytes, stem cells, fibroblasts, and tumor cells, the regulation and composition of the migrosome shows cell type-specific characteristics, and the old concept of 'one protein, one function' has to be revisited with regard to this cell function.

Cell migration is the crown achievement in biology. Cell locomotion is the most easily visible and yet one of the most complex processes exhibited by a living cell – complex because numerous cell surface molecules, macromolecules and organelles are implicated and the entire cell is involved. All molecules which are required for the various aspects of cell migration as major players within the migratory machinery can be coined as the cellular migrosome. The migrosome can timely differentially exist by an elaborated network of multi-protein complexes consisting of adhesion receptors, cytoskeletal components, signaling molecules and diverse adaptor proteins. The migrosome is formed mostly by a large cluster of transmembrane receptors of focal adhesion proteins, including sensory functional and signal transduction proteins to center local mechanical forces in orchestrating their complex interplay between the extracellular matrix and the dynamic cytoskeleton of a crawling cell.

Cell movement is the crown achievement in biology, because without cellular locomotion there is no life. Signalling of the female reproductive tract is central to regulate sperm motility, the majority of male infertility results from poor sperm motility. Parietal endoderm contributes to the yolk sac and is the first migratory cell type in the mammalian embryo. It has been shown recently that the parietal endoderm migration is directed by the non-canonical Wnt planar cell polarity pathway via Rho/ROCK [1]. ROCK inhibition leads to increased and diseased cell migration because these cells lack oriented migration. Directional cell migration is essential for almost all organisms during embryonic development, in adult life impairment contributes to pathological conditions. During embryogenesis it is essential that cells end up in their correct, precise locations in order to build a normal embryo. Hematopoietic and mesenchymal stem cell migration together with the ability to perceive and to percept the correct 'go' and 'stop' signals – migrosome-centered – are prerequisites to promote tissue repair and regeneration of the body. Stem cell therapy, e.g. in hematological disorders, might fail if the molecular processes which underlie the mobilization and directed migration from bone marrow into the peripheral tissues and back to the bone marrow compartment are disrupted. Mesenchymal stem cells are also multipotent cells which can support hematopoiesis, have immunomodulatory properties, may differentiate into osteocytes, chondrocytes and adipocytes, and specifically migrate to damage sites. The mesenchymal stem cell migration is mediated by growth factors, chemokines, adhesion molecules and toll-like receptors. Understanding the fundamental mechanisms underlying mesenchymal stem cell migration holds the promise of developing novel clinical strategies in regenerative medicine.

Wound healing requires fibroblast and keratinocyte migration. Non-healing cutaneous wounds, a major cause of morbidity and mortality, are difficult to treat. Recent studies suggest that significant increases in skin wound healing occur by altering gap junction intercellular communication. Gap junction intercellular communication can directly influence keratinocyte and fibroblast migration [2], and diseased migration can dramatically hamper wound closure.

The idea of relating cancer to stem cells is increasingly popular due to the identification of specific cancer stem cells sharing the typical plasticity and motility of pluripotent stem cells. Tumor invasion is driven by proliferation and importantly migration into the surrounding tissue. Cancer cell motility is also critical in the formation of metastases and is therefore a fundamental issue in cancer research. In solid tumors, the pivotal questions is from which tumor cells within a tumor mass a motile invasive phenotype emerges within a wide range of intratumoral microenvironmental growth conditions. Evidence has accumulated indicating that only a minority of cancer cells with stem cell properties, cancer stem cells (cancer stem-like cells), are responsible for maintenance, growth and metastases formation of tumors [3]. However, whether cancer stem-like cells give rise to metastases formation remains elusive despite vast information on cancer cells. Recently, Pawelek and Chakraborty [4] suggested that cancer cell fusion with macrophages or other migratory bone marrow-derived cells provides an explanation, because hybrids express mesodermal traits and epithelial-mesenchymal transition regulators (Twist, SPARC). Therefore, tumor-infiltrating immune cells are Janus-faced; either they form a cytotoxic T lymphocyte or macrophage immunological synapse to kill a tumor cell or they fuse with tumor/cancer stem-like cells generating a motile and invasive phenotype [5].

Cells migrate in embryogenesis to shape tissues, to vascularize tissues, in wound healing and most importantly, to exhibit immune competence. Immune competent cells have to patrol through the body, getting instructed within the lymph nodes for humoral and cellular attack formations to clear the body from foreign invaders and, hopefully, tumor cells. Immune competence is not only broken by inducing tolerance via blocked antigen recognition mechanisms but also by arresting the migratory machinery of immune competent cells. Immune competent cells mostly migrate as individuals, whereas tumor cells migrate more collectively in tightly or loosely associated clusters.

Cell movement is the crown achievement and essential for any living organism in order to leave hostile places and find an appropriate environment or food. Thus, movement makes only sense when the moving cell or organism is able to perceive where to move. This requires the ability to recognize a source or higher concentration of an attractive substance (e.g. food or pheromones) or physical condition (e.g. light or temperature). With regard to a single moving cell, such a perception has to be translated by an intracellular signal transduction, and to be coupled to the cytoskeleton and the locomotory machinery of the cell. Therefore, the cell migration process generally consists of three parts: the initiation of migration by the cognition of a stimulus by receptors, the intracellular transduction of the signal, and the dynamic orientation and movement of the cell into the right direction. From prokaryotes to eukaryotes and metazoans, evolution has on the molecular level found many ways to achieve this, whereas within the eukaryotes from fungi to mammalians there are common elements. Therefore, research on the fungus *Dictyostelium discoidium* is an eligible model even for the understanding of the migration of cells within a mammalian organism; the principal elements of sensing, recognition and chemotaxis are the same.

Back in 1976, Goldman et al. [6] edited three books on the mechanisms of cell motility. They comprehensively described the function and regulation of contractile molecules and motor proteins from the bacterial flagella to mammalian myosins. In this book, we focus on the current knowledge and the gain of knowledge since 1976 of the physiological and pathophysiological migration of autonomous cells in adult human organisms. After embryonic development, the ability to migrate is shut down in most of the differentiated cells, although the cells still express locomotor proteins such as actin and myosin. Some cells are even able to contract or change their morphology, however they are not able to migrate autonomously since they are not able to survive when unhinged from the united cell structure. Therefore, solely stem cells, leukocytes and fibroblasts constitute the fraction of physiologically and autonomously migrating cells in an adult human organism. On the opposite, tumor cells retrieve the ability to migrate during tumor progression, i.e. invasion and metastasis formation, and these cells are able to leave and survive without the united cell structure. As presented in this book, an intensive discussion is in progress on whether and how stem cells are involved in cancer development, putting forward the theory of cancer stem cells.

As discussed above, the migrating cells need to sense where to go. The trafficking within the body is regulated by ligand-receptor interactions, whereas the guidance molecules need to occur in a gradient that allows chemotactical movement towards the higher concentration. It is largely accepted that G-protein-coupled receptors (GPCRs) constitute the most important family of regulatory receptors for chemotactical movement. Chemokines and neurotransmitters are important groups of GPCR ligands, and especially chemokines regulate the localization of leukocytes at primary and secondary lymph organs as well as at sites of inflammation and injury. This regulatory function is addressed in several chapters of this book. Besides GPCRs, receptor tyrosine kinases are found to regulate cell migration too, which are activated by growth factors and cytokines. However, not only the physiologically migrating cells, but also tumor cells seem to underlie traffic and localization signals. Oncologists are well aware that certain types of cancer follow distinct patterns of metastasis formation depending on their tissue origin. Stephen Paget [7] was in 1889 the first who formulated the hypothesis that the spread of cancer cells, which he called the seed, can only grow to metastases in tissues of a certain constitution, which he called the soil. Today, we know that tumor cells are generally able to recognize gradients of chemoattractive substances and respond accordingly [8]. Furthermore, it has been shown in vivo that the CXC chemokine SDF-1 (stromal cell derived factor-1) functions as localization signal for metastases of breast and renal cancer [9, 10].

The directional movement requires a polarization of the cells into a front and rear end. We are at the beginning to understand how this is accomplished in cells which are exposed to a chemotactical gradient, and how this extracellular gradient is reflected in an intracellular organization. However, most in vitro migration assays do not work with gradients, and the induction of migratory activity by the addition of signal substances has therefore rather to be termed chemokinesis than chemotaxis. It is still unclear how the cells start to polarize and migrate in a random fashion without sensing an aim to move to. However, it was very recently shown that the coordinated regulation of pseudopod generation, orientation and persistence by multiple signaling pathways allows eukaryotic cells to detect extremely shallow gradients [11]. It could be shown that a genetically encoded photoactivatable Rac controls the motility of living cells. Localized Rac activation or inactivation was sufficient to produce cell motility and control the direction of cell movement. Myosin was involved in Rac control of directionality but not in Rac-induced protrusion, whereas PAK was required for Rac-induced protrusion [12]. The susceptibility of tumor cells in respect to fuel the migratory machinery by neurotransmitter was already presented in a monography in 2007 [13]. The overall effect of norepinephrine on the regulation of cancer cell migration and invasion and the blocking effect of propanolol using a model of pancreatic cancer cell lines – Miapaca-2 and Bxpc-3 – was recently confirmed by Guo et al. [14].

It is an experimental and clinical experience that tumors which are repopulated after treatment with chemo- and/or radiotherapy show an increased malignancy which is reflected by an increased capacity of proliferation, invasiveness and metastases formation. A Japanese group has very recently shown that cell migration, adhesion and invasion are enhanced in radiation-surviving cells, using the non-small cell lung cancer cell line H1299 [15] as a model cell line. Cells which survived radiation adhered more tightly to collagen-coated dished than parental non-irradiated cells. Interestingly, molecules – paxillin, phosphorylated FAK, integrin β_1 and vinculin – which are focally organized within the migrosome could be clearly visualized as key molecules for migratory, and in addition with increased expression levels of matrix metalloproteinases – MMP1, MMP2, and MMP9 – for invasive activities. On the other hand, the transcription factors HOXA4 and NOTCH3 might become interesting targets for inhibition of tumor cell migration, because both are involved in orchestrating the pathways which are responsible to ignite cell motility [16, 17].

The ultimate goal of cell migration research will be to understand and describe the signals of cell motility induction, the signal transduction pathways and the proteins, which are key elements responsible for the dynamic changes of the cytoskeleton. The old concept of 'one protein, one function' is no longer adequate and it is necessary to conceive an ensemble of proteins, focused and task-orientated within a migrosome, thereby using many different pathways, either simultaneously or interchangeably, to implement the crown achievement of biology in an organism, namely cell migration.

Acknowledgement

This work was supported by the Fritz Bender Foundation (Munich, Germany).

References

1 LaMonica K, Bass M, Grabel L: The planar cell polarity pathway directs parietal endoderm migration. Dev Biol 2009;330:44–53.

2 Wright CS, van Steensel MA, Hodgins MB, Martin PE: Connexin mimetic peptides improve cell migration rates of human epidermal keratinocytes and dermal fibroblasts in vitro. Wound Repair Regen 2009;17:240–249.

3 Dittmar T, Zänker KS: Cancer and Stem Cells. New York, Nova Science, 2008.

4 Pawelek JM, Chakraborty AK: The cancer cell-leukocyte fusion theory of metastasis. Adv Cancer Res 2008;101:397–444.

5 Dittmar T, Nagler C, Schwitalla S, Reith G, Niggemann B, Zänker KS: Recurrence cancer stem cells – made by cell fusion? Med Hypotheses 2009;73:542–547.

6 Goldman R, Pollard T, Rosenbaum J: Cell Motility. Cold Spring Harbor, Cold Spring Harbor Laboratory, 1976.

7 Paget S: Distribution of secondary growths in cancer of the breast. Lancet 1889;133:571–573.

8 Bastian P, Posch B, Lang K, Niggemann B, Zaenker KS, Hatt H, Entschladen F: Phosphatidylinositol 3-kinase in the G-protein-coupled receptor-induced chemokinesis and chemotaxis of MDA-MB-468 breast carcinoma cells: a comparison with leukocytes. Mol Cancer Res 2006;4:411–421.

9 Muller A, Homey B, Soto H, Ge N, Catron D, Buchanan ME, McClanahan T, Murphy E, Yuan W, Wagner SN, Barrera JL, Mohar A, Verastegui E, Zlotnik A: Involvement of chemokine receptors in breast cancer metastasis. Nature 2001;410:50–56.

10 Pan J, Mestas J, Burdick MD, Phillips RJ, Thomas GV, Reckamp K, Belperio JA, Strieter RM: Stromal derived factor-1 (SDF-1/CXCL12) and CXCR4 in renal cell carcinoma metastasis. Mol Cancer 2006;5:56.

11 Bosgraaf L, Van Haastert PJ: Navigation of chemotactic cells by parallel signalling to pseudopod persistence and orientation. PLoS One 2009;4:e6842.

12 Wu YI, Frey D, Lungu OI, Jaehrig A, Schlichting I, Kuhlmann B, Hahn KM: A genetically encoded photoactivatable Rac controls the motility of living cells. Nature 2009;461:104–108.

13 Zänker KS, Entschladen F: Neuronal activity in tumor tissue; in Bertino JR (ed): Prog Exp Tumor Res. Basel, Karger, 2007, vol 39.

14 Guo K, Ma Q, Wang L, Hu H, Li J, Zhang D, Zhang M: Norepinephrine-induced invasion by pancreatic cancer cells is inhibited by propanolol. Oncol Rep 2009;22:825–830.

15 Tsutsumi K, Tsuda M, Yazawa N, Nakamura H, Ishihara S, Haga H, Yasuda M, Yamazaki R, Shirato H, Kawaguchi H, Nishioka T, Ohba Y: Increased motility and invasiveness in tumor cells that survive 10 Gy irradiation. Cell Struct Funct 2009;34:89–96.

16 Klausen C, Leung PC, Auersperg N: Cell motility and spreading are suppressed by HOXA4 in ovarian cancer cells. Possible involvement of β_1 integrin. Mol Cancer Res 2009;7:1425–1437.

17 Song G, Zhang Y, Wang L: MICRORNA-206 targets NOTCH3, activates apoptosis, inhibits tumor cell migration and foci formation. J Biol Chem 2009; epub ahead of print, PMID 19723635.

Prof. Dr. Frank Entschladen
Institute of Immunology, Witten/Herdecke University
Stockumer Strasse 10, DE–58448 Witten (Germany)
Tel. +49 2302 926 187, Fax +49 2302 926 158, E-Mail entschladen@uni-wh.de

Entschladen F, Zänker KS (eds): Cell Migration: Signalling and Mechanisms.
Transl Res Biomed. Basel, Karger, 2010, vol 2, pp 7–27

Stem Cell Migration in Health and Disease

Thomas Dittmar[a] · Susannah H. Kassmer[d] · Benjamin Kasenda[b] · Jeanette Seidel[c] · Bernd Niggemann[a] · Kurt S. Zänker[a]

[a]Institute of Immunology, Witten/Herdecke University, Witten; [b]Department of Hematology and Oncology, University of Freiburg Medical Center, Freiburg, and [c]Medizinische Klinik II m. S. Hämatologie/Onkologie, Charité Campus Mitte, Berlin, Germany, and [d]Department of Laboratory Medicine, Yale Stem Cell Center, Yale University, New Haven, Conn., USA

Abstract

Within the past years, our knowledge about stem cells in health and disease has changed dramatically. To date, it is feasible to isolate and propagate human pluripotent stem cells from various sources, such as cord blood, bone marrow or adipose tissue, and to generate donor-specific ethically harmless induced pluripotent stem cells, which exhibits embryonic stem cell properties. However, irrespective of the used stem cell type(s), the success of tissue regeneration therapies does not only depend on the cells' differentiation capacity, but also on their ability to migrate. Without migration, stem cells would neither be able to reach the appropriate degenerated tissue (if administered intravenously) nor they would be able to regenerate it because restoration of organ tissue integrity and function means to reconstruct a three-dimensional organ environment. However, the ability of stem cells to migrate is not only crucial for tissue regeneration processes, but do also play a role in tumor progression. To date, we know that cancer has its origin in a small subpopulation of cancer cells exhibiting stem cell properties, the so-called cancer stem cells. Because of their tumor initiation capacity, cancer stem cells have now also been linked to metastasis formation, which prerequisites cell migration. In summary, stem cell migration is a crucial process in both health and disease.

Within the past years, our knowledge about stem cells in health and disease has changed dramatically. Today we know that stem cells from bone marrow or from adipose tissue are pluripotent and that they can be used for regenerative purposes. Likewise, the creation of donor-specific stem cells (so-called induced pluripotent stem cells (iPS cells), exhibiting embryonic stem cell (ESC) properties) simply by transducing two to four transcription factors sounded nearly unbelievable a couple of years ago.

However, which type(s) of stem cell(s) will be ultimately used for regenerative medical purposes is not yet clear because each stem cell type has its own pros and cons. For

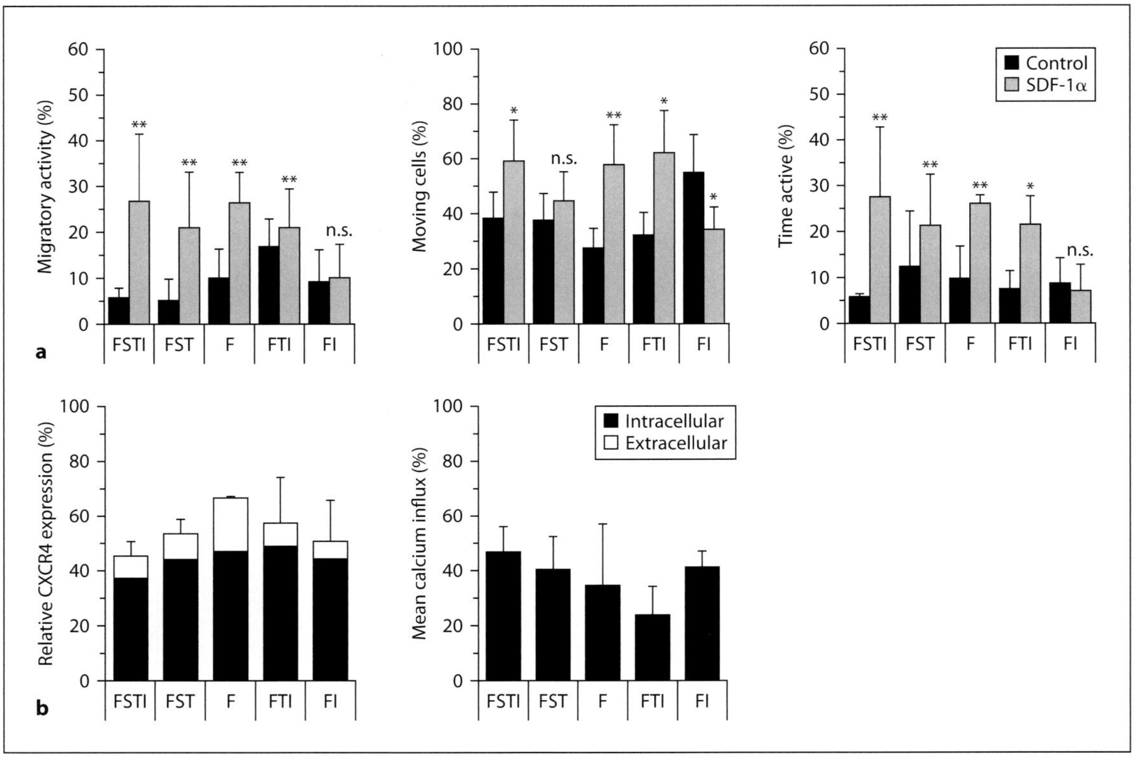

Fig. 1. Influence of culture conditions on the migratory phenotype of murine Lin⁻ c-kit⁺ HSPCs. Murine Lin⁻ c-kit⁺ HSPCs were cultivated for 5 days in the presence of various combinations of Flt3-ligand (F), SCF (S), TPO (T), and IL-11 (I) prior to analysis of their migratory activity in dependence of SDF-1α stimulation. **a** Migration pattern of cultured murine Lin⁻ c-kit⁺ HSPCs show that each cytokine/cytokine combination had a distinct effect. For instance, FSTI-cultured cells responded very well to SDF-1α stimulation, whereas FI-cultivated cells did not show an increased migratory activity in response to SDF-1α treatment. Cell migration parameters (moving cells, time active) indicated that SDF-1α rather acts as an inhibitor on FI-cultured cells. **b** Comparable CXCR4 expression levels were detectable on cultured cells. Likewise, increased cytosolic calcium concentrations were detectable after SDF-1α stimulation in all culture cells indicating that the engagement of CXCR4-specific signal transduction cascades. Shown are representative data of 5 out of 12 cytokine/cytokine combinations. Statistical significance (paired Student's t test): n.s. = not significant, * $p < 0.01$, ** $p < 0.001$.

instance, most adult stem cells, such as hematopoietic stem/progenitor cells (HSPCs), do not remain in a stem cell state under in vitro conditions. HSPC cultivation for a week and longer is associated with a loss of the HSPC marker molecules CD34 and CD133, indicating induction of differentiation. Optimized culture conditions can delay HSPC differentiation in culture but may possess an unknown risk concerning the cells' ability to respond the chemokine stromal cell-derived factor-1α (SDF-1α) [1], which to date is still the most prominent chemoattractant for HSPCs [2]. We have recently demonstrated that murine Lin⁻ c-kit⁺ HSPCs, which have been cultured in the presence of Flt3-ligand and interleukin (IL)-11 for up to 5 days, did not respond to SDF-1α stimulation with an increased migratory activity (fig. 1a) [1]. Moreover, a detailed analysis of several migration parameters, the so-called migration pattern (fig. 1a) [1, 3, 4], revealed that SDF-1α rather acted as an inhibitor of cell migration on such cultured murine HSPCs [1]. On the other hand, most adult stem cells are easily accessible and can be used in an autologous manner, which avoids immunosuppression of treated patients. The latter is indispensable if ESCs will be used for regenerative purposes because it is not yet feasible to create donor-specific ESCs. ESCs remain in their stem cell state in vitro, can be propagated nearly unlimited and possess an unrestricted differentiation capacity in vitro and in vivo. On the other hand, human ESCs are still a subject to controversial and ethical discussions because of the way how these cells will be generated. Likewise, ESCs cannot be administered directly in degenerated tissues while this would result in teratoma formation. This problem can be overcome if ESCs will be predifferentiated prior to implantation, whereby predifferentiation is associated with an overall decreased survival rate of transplanted cells.

An alternative source of ethically harmless stem cells exhibiting ESC properties might be iPS cells, which can be generated from adult somatic cell or adult stem cells by transduction of two or four transcription factors [5, 6]. In fact, iPS cells possess several ESC characteristics including morphology, proliferation, gene expression, telomerase activity, epigenetic status, the capacity of unrestricted differentiation, and teratoma formation in vivo. The latter property of iPS cells is used as a read-out for true iPS cell generation. Quite recently, Zhou et al. [7] demonstrated the feasibility to generate iPS cells without the use of viral vectors or plasmids by using recombinant proteins capable of penetrating the plasma membrane of cells. Therefore, a polyarginine protein transduction domain was fused to the C-terminus of the four reprogramming factors Oct4, Sox2, KLf4, and c-Myc [7]. After expression in bacteria and purification, these recombinant proteins are simply added to iPS cell media and autonomously find their way into the cells, where they induce reprogramming. Although the efficacy of this iPS cell generation method is still rather low, this methodology overcomes a putative severe side effect of plasmids and particularly viral vectors: the random integration into the host genome, which bears potentially tumorigenic risks.

Irrespective of the pros and cons of the various stem cell types, they all have one parameter in common, which, besides the capacity to differentiate, will be crucial for successful tissue regeneration: the ability of migrate. Without this property, stem cells would not be able to regenerate degenerated tissues because the cells would not reach their final destination within a three-dimensional organ tissue environment. Even if stem cells were applied directly into the appropriate destructed tissues they would have to move into the surrounding periphery to reconstruct the three-dimensional organ architecture. Thus it is of crucial interest not only to investigate the differentiation and tissue restoration capacity of different stem cell types, but also their migratory behavior.

In the following chapter we will primarily focus on hematopoietic stem cells/hematopoietic stem/progenitor cells (HSCs/HSPCs) since the regulation of the migration of these stem cells is well characterized. Here, we will give an overview about the signals and molecules that direct HSC/HSPC migration during mobilization and homing. In addition to this, we will also give a short summary about the putative migratory capacities of a different type of stem cell, which has become much of interest in the past years, namely cancer stem cells (CSCs). It is now generally acknowledged that cancer originates from CSCs, because of their (tumor) tissue restoration capacity [8, 9]. If we agree with that then we can conclude that metastases should also arise from CSCs. In this context, Li et al. [10] postulated the existence of metastatic CSCs (mCSCs) representing a distinct type of CSCs that initiates secondary tumor growth. Here we will give a short summary about the current knowledge of mCSCs and the molecules that likely direct their migration.

HSC/HSPC Mobilization

HSCs/HSPCs reside within their specialized bone marrow niches (fig. 2) and give rise to more committed progenies, which successively differentiate in mature blood cells that migrate into the circulation. Under normal conditions a small proportion of primitive HSCs/HSPCs constantly leaves the bone marrow and circulates through the bloodstream. In relation to the total number of peripheral blood mononuclear cells (PBMCs), the amount of circulating HSPCs is about 0.05–0.1% [11], thus being rather small. However, the pool of circulating HSPCs can be significantly increased within the PBMC fraction to up to 3% via administration of cytokines and chemokines [12, 13]. This process, called mobilization, is characterized by both a loss of cell-to-cell contacts due to downregulation and degradation of cell adhesion molecules, and desensitization of the SDF-1α/CXCR4 axis (fig. 2) [13]. For most cytokines and chemokines the mechanism in detail how they induce mobilization remains unclear. This belongs as well to the phenomenon that HSC/HSPC mobilization mediated by cytokines, such as granulocyte-colony stimulating factor (G-CSF) and granulocyte-macrophage-colony stimulating factor (GM-CSF), generally requires 5–6 days for a peak level response, whereas chemokines, like IL-8 and growth-regulated oncogene-β (GROβ), induce mobilization within the time span of 30 min to a few hours [13].

Both G-CSF/GM-CSF as well as IL-8/GROβ activate neutrophil granulocytes to secrete various proteases, such as elastase and cathepsin G and matrix metalloproteinase-9 (MMP-9) [14–16], which facilitate mobilization by degradation of cell adhesion molecules and desensitization of the SDF-1α/CXCR4 axis (fig. 2). For instance, upon prolonged G-CSF treatment increased elastase and cathepsin G concentrations can be found within the bone marrow, which is correlated to a sharp reduction of vascular cell adhesion molecule-1 (VCAM-1/CD106) expression of bone marrow stroma cells [17]. VCAM-1/CD106 binds to very late antigen-4 (VLA-4; $\alpha_4\beta_1$-

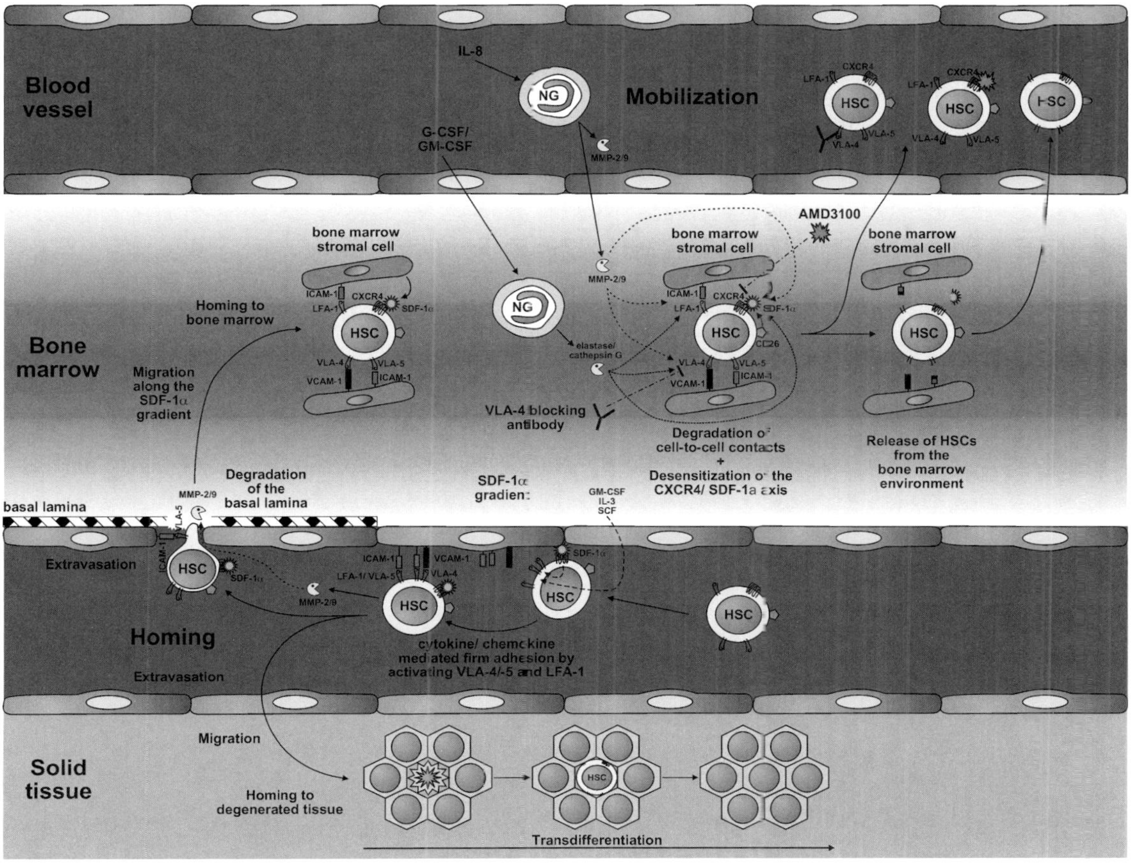

Fig. 2. Mobilization and homing of HSCs/HSPCs. HSC/HPSC mobilization and homing are mirror processes that strongly depend on the activation state of cell-to-cell contacts and the SDF-1α/CXCR4 axis. HSC/HSPC mobilization from the bone marrow mediated by cytokines (e.g. G-CSF) or chemokines (e.g. IL-8) is caused by proteases (MMPs, elastase and cathepsin G) that degrade various adhesion molecules as well as SDF-1α and its receptor CXCR4. Mobilization is also achieved by administration of a VLA-4 ($\alpha_4\beta_1$-integrin) blocking antibody or by application of the CXCR4 antagonist AMD3100. In contrast, upregulation of cell adhesion molecules and activation of the SDF-1α/CXCR4 axis is essential for stem cell homing to both bone marrow and solid tissues. Shown here in detail is the extravasation of HSCs/HSPCs into the bone marrow, which is similar to the extravasation of HSCs/HSPCs into solid tissues [reprinted from 13, with permission].

integrin), which is expressed by HSCs/HSPCs. Among other cell-to-cell contacts between HSCs and bone marrow stroma cells, e.g. mediated by intercellular adhesion molecule-1 (ICAM-1)/leukocyte function-associated antigen-1 (LFA-1) as well as ICAM-1/VLA-5 ($\alpha_5\beta_1$-integrin) [13], the interaction between VCAM-1/CD106 and VLA-4/$\alpha_4\beta_1$-integrin appears to play a crucial role in the retention of HSCs inside their niche (fig. 2). Recent studies revealed that the $\alpha_4\beta_1$-integrin-blocking antibody natalizumab, which was originally developed for the treatment of multiple sclerosis patients to prevent CD8 extravasation into the brain, significantly increased the number of circulating HSCs/HSPCs within the peripheral blood [18, 19].

Desensitization of the SDF-1α/CXCR4 axis mediated by proteases is chiefly facilitated by cleavage of the N-terminus of the SDF-1α receptor CXCR4 on HSCs [20]. The G-CSF-mediated mobilization of HSCs/HSPCs also involves the membrane-bound extracellular peptidase CD26 (dipeptidylpeptidase IV (DPPIV)), which is expressed on a subset of CD34+ HSCs [21]. CD26/DPPIV inactivates SDF-1α by cleaving it at its position two proline [21]. Treatment of mice with CD26 inhibitors during G-CSF-induced mobilization resulted in a reduced number of progenitor cells in the periphery as compared to the G-CSF regimen alone [22]. Thereby, G-CSF upregulates CD26/DPPIV expression and activity resulting in an enhanced SDF-1α degradation concomitant with desensitization of the SDF-1α/CXCR4 axis [23].

Desensitization of the SDF-1α/CXCR4 axis is also achieved by the bicyclam molecule AMD3100 (Plerixaflor, Genzyme Corp.) that antagonizes binding of SDF-1α to its receptor CXCR4 [24]. In combination with G-CSF, AMD3100 leads to the rapid mobilization of long-term repopulating HSCs [25]. AMD3100 (Plerixaflor, Genzyme Corp.) was approved by the Food and Drug Administration (FDA) in December 2008, thereby representing the first CXCR4 inhibitor that is used as a HSC/HSPC-mobilizing drug.

HSC/HSPC Homing

Homing, as defined by Lapidot [26], is a descriptive term used for the crossing of circulating HSCs/HSPCs across the blood/bone endothelial barrier into the bone marrow compartment within a fairly short time span of hours to days. Successful homing is measured by the successful reconstitution of hematopoiesis, whereby successful tissue restoration and organ function could also be used as a read-out for successful homing [27]. HSC/HSPC homing is a multistep process requiring the interplay of adhesion molecules, cytokines, chemokines, and extracellular matrix-degrading proteases. It thus resembles leukocyte/lymphocyte [28] as well as tumor cells extravasation during hematogenous metastatic spreading [29] and can be subdivided into the rolling phase, the adhesion phase, and transendothelial migration. The rolling phase of extravasation of HSCs/HSPCs is mediated by E- and P-selectins [30, 31], whereas the firm adhesion is facilitated through ICAM-1/LFA-1, VCAM-1 (CD106)/VLA-4 ($\alpha_4\beta_1$-integrin) interactions [32–34]. Likewise, transendothelial migration of HSCs/HSPCs is mediated by PECAM-1 (CD31) [35].

In addition to its role in facilitating the retention of HSCs in their specific niche, the SDF-1α/CXCR4 axis is also mandatory for the homing process. On the one hand, successful homing depends on the active migration of along a SDF-1α gradient [2, 36, 37]. On the other hand, SDF-1α triggers the firm adhesion of HSCs/HSPCs to the endothelium by activating the integrins LFA-1, VLA-4, and VLA-5 on hematopoietic cells under shear flow conditions (fig. 2) [32, 33, 38]. Likewise, SDF-1α induces the polarization and extravasation of HSCs/HSPCs via a both VLA-4- and VLA-5-dependent mechanism [33]. Polarization per se is a prerequisite for cell migration [39], which has been shown for various cell types including HSCs/HSPCs, leukocytes/lymphocytes, and tumor cells. In KG1a cells, which have been stably transfected with CXCR4, SDF-1α caused a relocalization of CXCR4 to the leading edge of transfected KG1a cell upon contact with human umbilical vein endothelial cells [40]. Thereby, CXCR4 is co-localized with lipid rafts and is found at cell-to-cell interaction sites when in contact with the endothelial cell surface [40].

Adhesion of HSCs/HSPCs to the endothelium during homing is further triggered by various cytokines. For instance, GM-CSF, IL-3, and SCF temporarily increase the adhesiveness of HSCs/HSPCs by activating the adhesion molecules VLA-4 and VLA-5 (fig. 2). Likewise, Flt3-ligand, SCF, IL-3, IL-6, and HGF upregulate CXCR4 expression on HSCs/HSPCs in vitro and in vivo [38, 41, 42], thereby enhancing the intracellular signals generated through the SDF-1α/CXCR4 axis [43]. Interestingly, prolonged cultivation of HSCs/HSPCs is associated with CXCR4 downregulation, whereby decreased SDF-1α receptor levels do not correlate to the cells' migratory activity in response to SDF-1α stimulation [1, 3, 4]. For instance, murine Lin$^-$ c-kit$^+$ HSPCs, which were cultivated for 5 days in the presence of Flt3-ligand, SCF, TPO, and IL-11 responded to SDF-1α stimulation with a markedly increased migratory activity [1]. By contrast, murine Lin$^-$ c-kit$^-$ HSPCs cultivated with a combination of Flt3-ligand and IL-11 did not respond to SDF-1α stimulation with an increased locomotory activity although CXCR4 expression levels (both intracellular and plasma membrane-bound) were comparable to Flt3-ligand-, SCF-, TPO-, and IL-11-cultured cells (fig. 1b) [1]. Moreover, SDF-1α stimulation of Flt3-ligand- and IL-11-cultured murine Lin$^-$ c-kit$^+$ HSPCs led to increased cytosolic calcium levels as well as activation of CXCR4-specific signal transduction cascades (fig. 1b) [1].

Cytokines do also trigger the transendothelial migration of HSCs/HSPCs by inducing the expression of matrix-degrading enzymes, like MMPs, which is mandatory for the degradation of the basement membrane [44]. Two recent studies by Zheng et al. [45, 46] showed that cord blood CD34+ HSCs exhibited significantly lower expression levels of CD49e, CD49f, CXCR4, MMP-2, and MMP-9, as compared to CD34+ HSC from bone marrow or peripheral blood. Upon SCF stimulation, cord blood CD34+ HSCs gained increased expression levels of CXCR4, MMP-2, MMP-9, and other homing-related molecules concomitant with an increased in vitro transendothelial migration capacity and in vivo homing potential [45, 46].

In addition to HSC/HSPC homing to the bone marrow, the SDF-1α/CXCR4 axis does also play a crucial role in directing circulating HSCs/HSPCs from the peripheral blood into degenerated tissues. Kollet et al. [41] were able to show that both irradiation and inflammation led to elevated SDF-1α expression levels in the liver bile and duct epithelial cells. Conjointly, hepatic injury induced MMP-9 activity leading to both increased CXCR4 expression levels and SDF-1α-mediated recruitment of HSCs/

HSPCs to the liver, whereby recruited cells were found in close proximity to SDF-1α expressing epithelial cells [41]. A putative function for the SDF-1α/CXCR4 axis in inflammatory conditions was already published by Gonzalo et al. [47] who demonstrated that SDF-1α is a critical inflammatory component in allergic airway disease. Inhibition of CXCR4 function by application of CXCR4-neutralizing antibodies in a mouse model of allergic airway disease resulted in reduced eosinophilia by half concomitant with a significant decrease in airway hyperresponsiveness [47]. Likewise, elevated SDF-1α levels have also been associated with the recruitment and accumulation of CD4+ lymphocytes in rheumatoid arthritis synovium [48, 49]. Thereby, CXCR4 expression on CD4+ T cells was increased by both IL-15 and TGF-β, whereas SDF-1α expression of the synovium and of synovial endothelial cells was increased by CD40 stimulation [48, 49] suggesting a self-energizing feedback loop.

The role of the SDF-1α/CXCR4 axis in inflammatory conditions is further supported by findings of Ceradini and Gurtner [50] showing that the recruitment of CXCR4-positive progenitor cells to regenerating tissues is mediated by the hypoxic gradient, namely via HIF-1-induced expression of SDF-1α in endothelial cells. Thereby, the upregulation of SDF-1α in ischemic tissue is directly proportional to reduced oxygen tension, as well as it is correlated to an increased adhesion, migration, and homing of circulating HSCs/HSPCs to ischemic tissue [50].

The SDF-1α/CXCR4 Axis

The interaction between SDF-1α and its receptor CXCR4 (also named the SDF-1α/CXCR4 axis) plays a pivotal role in regulating the retention, migration, mobilization, and homing of HSCs/HSPCs during steady-state homeostasis and tissue injury [51]. In addition to HSCs/HSPCs, activation of the SDF-1α/CXCR4 axis also initiates the migration of lymphocytes [52]. Moreover, within the past year it became evident that the progression and organ-specific metastatic spreading of various cancer types was associated with the SDF-1α/CXCR4 axis [53, 54], which acts a navigation system for circulating tumor cells [29].

Cell migration is an essential component of both successful mobilization and homing of HSCs/HSPCs. However, cell migration is a complex process, which is directed by the interplay of several signal transduction pathways initiated by various ligands, such as cytokines, chemokines, and extracellular matrix components that activate growth factor receptors, chemokine receptors and integrins [55].

The SDF-1α-induced chemotaxis of HSCs/HSPCs is inhibited by pertussis toxin, indicating that CXCR4 is associated with a $G\alpha_i$-protein subtype [2, 56, 57]. Binding of SDF-1α to CXCR4 activates several signal transduction cascades including the PI3-kinase (PI3K)/Akt pathway, the phospholipase C-γ (PLC-γ)/protein kinase C (PKC) pathway and the $MAPK^{p42/44}$ (ERK-1/2) pathway [58, 59]. Studies on human T-cell lines indicated that SDF-1α triggers CXCR4 dimerization and activates the

JAK/STAT pathway, which suggests gene regulation [60]. Likewise, Ganju et al. [58] reported that SDF-1α treatment led to increased NF-κB activity in nuclear extracts of CXCR4 transfectants, indicating that changes of the gene expression level can be initiated via two independent signal transduction pathways downstream of the SDF-1α/CXCR4 axis. By contrast, in factor-dependent MO7e cells, NF-κB did not appear to be involved in SDF-1α actions [43].

The actin cytoskeleton is one of the central mechanical components responsible for the motility of cells, and its analysis is an effective method of determining a migratory phenotype. Actin polymerization in migration is induced by the PI3K/Akt signaling and the PLC-γ/PKC cascade in a variety of cell types [61–64], including HSCs/HSPCs. So far, several groups have convincingly demonstrated that SDF-1α induces actin polymerization [2, 65] as well as tyrosine phosphorylation of several components of focal adhesion complexes such as paxillin, the related adhesion focal tyrosine kinase (RAFTK/pyk2), $p130^{cas}$, Crk-II, and Crk-L [58, 59]. Inhibition of PI3K signaling using wortmannin partially inhibited the SDF-1α-induced migration and tyrosine phosphorylation of paxillin [58], further underpinning the role of PI3K/Akt signaling in HSC/HSPC migration.

In a recent study by Petit et al. [66], the SDF-1α-mediated cell polarization, adhesion to bone marrow stromal cells, and chemotaxis of human CD34+ progenitor cells were all shown to be PKC-ζ-dependent. PKC-ζ belongs to the group of atypical PKC isoforms in which activation does not depend on calcium or diacylglycerol [67, 68]. Petit et al. [66] identified PI3K as an activator of PKC-ζ, and Pyk-2 and $MAPK^{p42/44}$ (ERK-1/2) as downstream targets of PKC-ζ [66]. In vivo studies showed that the engraftment, but not homing, of human CD34+ HSPCs was also PKC-ζ-dependent.

In contrast to this we have recently demonstrated that the migratory activity of both Flt3-ligand- and Flt3-ligand/IL-6-cultivated cord blood CD34+/CD133+ HSPCs was markedly inhibited by Gö6976 (a specific PKC-α inhibitor [69]) treatment on day 1 (fig. 3a) suggesting an involvement of PKC-α in CD34+/CD133+ HSPC migration [3]. Interestingly, analysis of the locomotory behavior of cord blood CD34+/CD133+ HSPCs cultured for 5 days with either Flt3-ligand alone or in combination of Flt3-ligand/IL-6 revealed a different migratory phenotype. For solely Flt3-ligand-cultured cord blood CD34+/CD133+ HSPCs, we noticed only a slightly decreased migratory activity in the presence of the PKC-α inhibitor Gö6976, whereas for day 5 Flt3-ligand/IL-6-cultured cord blood CD34+/CD133+ HSPCs no inhibitory effect of Gö6976 was observed (fig. 3a) [3]. In fact, the migratory activities of untreated and Gö6976-treated as well as SDF-1α- and SDF-1α/Gö6976-treated Flt3-ligand/IL-6-cultured cord blood CD34+/CD133+ HSPCs were virtually identical. Western blot analysis revealed comparable PKC-α expression levels in both Flt3-ligand- and Flt3-ligand/IL-6-cultured cord blood CD34+/CD133+ HSPCs on both day 1 and day 5 (fig. 3b) [3]. These findings indicate that the involvement of PKC-α in the process of cell migration is altered during the prolonged culture period of 5 days. Although PKC-α expression is clearly detectable in both Flt3-ligand- and Flt3-ligand/IL-6-cultured cord blood CD34+/

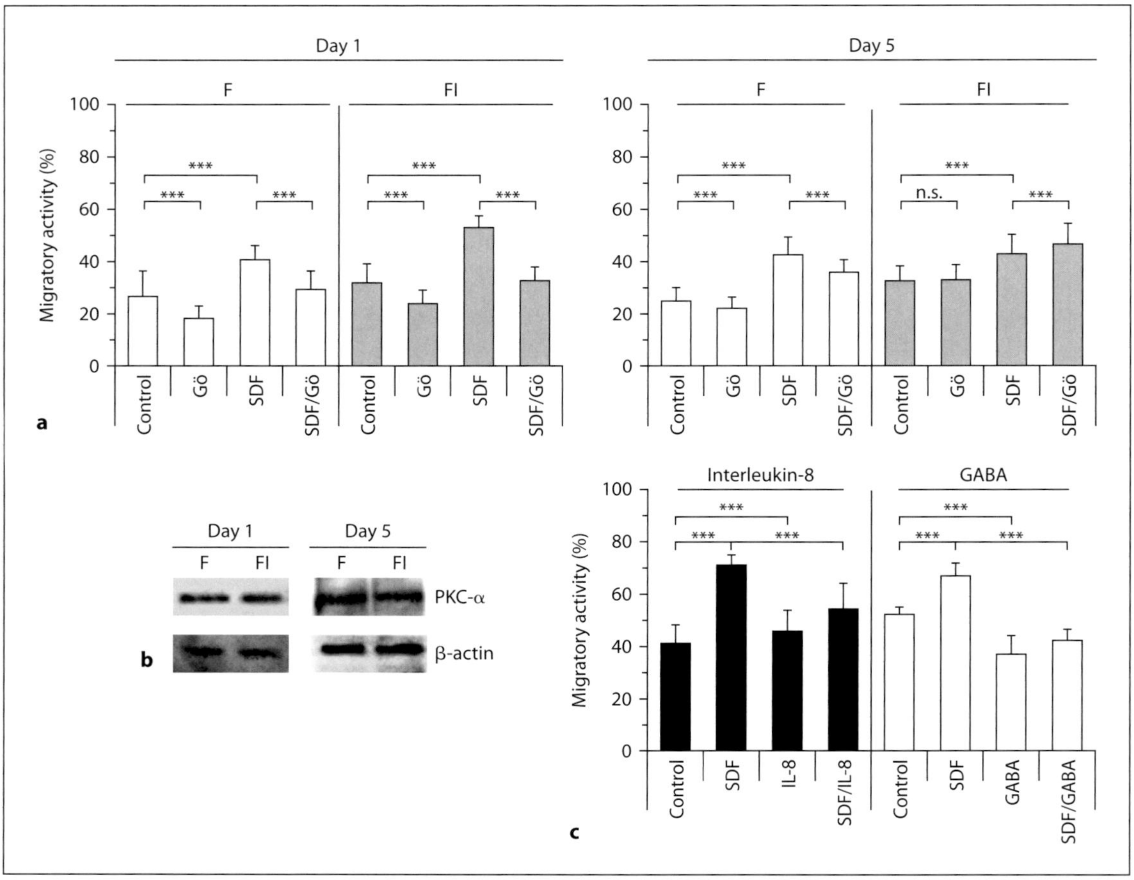

Fig. 3. Modulation of HSPC migration. The SDF-1α-dependent migration of HSCs/HSPCs is modulated by both culture conditions as well as soluble factors, such as IL-8 and GABA. **a** Cell migration data for cord blood CD34+/CD133+ HSCs/HSPCs that were cultivated for 1 and 5 days in the presence of Flt3-ligand (F) and a combination of Flt3-ligand and IL-6 (FI). Day 1 data show that both the spontaneous as well as the SDF-1α-dependent migratory activity of the cells could be blocked by the specific PKC-α inhibitor Gö6976 (Gö) suggesting a PKC-α-dependent migration. By contrast, day 5 data reveal solely a weak inhibition of cell migration by Gö6976 suggesting that a switch from a PKC-α-dependent towards a PKC-α-independent migratory phenotype might have occurred. Flt3-ligand (F) cultured cells appear in white, whereas Flt3-ligand/IL-6 (FI)-cultivated cells appear in gray. **b** PKC-α expression levels of Flt3-ligand (F)- and Flt3-ligand/IL-6 (FI)-cultured cells in relation to β-actin control revealed no differences between culture conditions and time of cultivation. **c** IL-8 and GABA impair the SDF-1α-induced migration of human adult CD34+/CD133+ HSCs/HSPCs. Cell migration data for IL-8 appear in black, whereas cell migration data for GABA are shown in white. Shown are the means of at least three experiments. Statistical significance (paired Student's t test): n.s. = not significant, * $p < 0.05$, ** $p < 0.01$, *** $p < 0.001$.

CD133+ HSPCs after 5 days of culture, it seems that this molecule not longer plays a key role in the migratory activity of these cells.

Modulation of the SDF-1α-Induced Migration of HSPCs

In most cell migration studies, factors and their induced signal transduction cascades are investigated that *induce* cellular movement. While the knowledge about that is definitely of interest for several purposes, e.g. to understand the molecular processes that direct the migration of different cell types, including HSCs/HSPCs [1, 3, 70], leukocytes/lymphocytes [64, 71], and tumor cells [29, 71], the question has to be addressed how the migration of the cells is terminated once they have reached their final destination. A couple of years ago, Lang et al. [72] reported that IL-8, which is generally known to be a potent chemoattractant for various cells, such as neutrophil granulocytes, is an inhibitor of cell migration. Thereby, IL-8 dose-dependently increased the frequency and the duration of stop periods of formyl-methionyl-leucyl-phenylalanine (fMLP)-induced neutrophil granulocytes [72]. Because IL-8 mediates HSC/HSPC migration [73] the question arose whether IL-8 in conjunction with SDF-1α might also inhibit the locomotory activity of adult CD34+/CD133+ HSPCs. Both IL-8 and SDF-1α alone stimulated the migratory activity of adult CD34+/CD133+ HSPCs, whereby the IL-8-mediated induction of cell migration was rather moderate, but significant (control: 41.5 ± 6.9% vs. 20 ng/ml IL-8: 46.1 ± 8.0%; fig. 3c). By contrast, the SDF-1α-induced migration of adult CD34+/CD133+ HSPCs was nearly doubled as compared to untreated control cells (control: 41.5 ± 6.9% vs. 1 μg/ml SDF-1α: 71.1 ± 3.7%; fig. 3c). Combination of IL-8 and SDF-1α yielded in a mean locomotory activity of about 55.0 ± 9.2%, which was in between the IL-8- and SDF-1α-induced locomotory activities. Whether IL-8 inhibits the SDF-1α-induced migration of adult CD34+/CD133+ HSPCs by a similar mechanism as IL-8 blocks the fMLP-induced migration of neutrophil granulocytes is not yet clear. Nonetheless, these data show that IL-8 is capable of impairing the SDF-1α-induced migration of adult CD34+/CD133+ HSPCs, which might play a role in the termination of these cells.

In addition to IL-8, the migration of adult CD34+/CD133+ HSPCs was also markedly blocked by the neurotransmitter γ-aminobutyric acid (GABA) (fig. 3c) [70]. To date, considerably less is known about GABA receptor expression and function in non-neuronal tissues. However, in a recent study, Rane et al. [74] demonstrated that $GABA_B$ receptors stimulate the chemotaxis of neutrophil granulocytes via PI3K/Akt signaling during ischemia reperfusion. In accordance with neutrophil granulocytes, $GABA_B$-receptor expression was also detected on adult CD34+/CD133+ HSPCs [70, 75], whereas GABA markedly blocked both the spontaneous and SDF-1α-induced migration of these cells (control: 52.5 ± 2.8%; 1 μg/ml SDF-1α: 67.3 ± 4.7%; 100 μM GABA: 37.4 ± 7.0%; 1 μg/ml SDF-1α + 100 μM GABA: 42.5 ± 4.2%; fig. 3c) [70]. Mechanistically, GABA most likely impairs cell migration by inhibiting calcium release-activated calcium (CRAC) channels [70]. However, whether $GABA_B$-receptor signaling directly interacts with CRAC channels, thereby inhibiting it or whether $GABA_B$-receptor signaling indirectly impairs CRAC channel function due to interaction with the CXCR4 signal transduction cascade remains unknown.

In summary, these data show that factors exist that inhibit the SDF-1α-induced migration of adult CD34+/CD133+ HSPCs. IL-8 is a well-known chemokine which recruits leukocytes to areas of inflammation. However, in combination with other compounds, such as fMLP and SDF-1α, IL-8 has an inhibitory effect on locomoting cells suggesting that IL-8 might function as a migration terminating factor, thereby regulating the motility of immunocompetent cells (and possibly HSPCs) in inflamed tissues. Likewise, the neurotransmitter GABA is a potent inhibitor of HSPC migration, whereby the role of GABA in non-neuronal tissues still remains ambiguous. However, the findings of Rane et al. [74] that GABA recruits neutrophil granulocytes suggests a putative role for GABA in inflammatory conditions.

Cancer Stem Cell Migration

CSCs have become much of interest within the past decade. CSCs represent a small population of cancer cells exhibiting stem cell properties, such as self-renewing, differentiation, tissue reconstitution and drug resistance [76–78]. Because of their tumor initiation capacity and resistance against cytotoxic drugs and radiation, CSCs have not only been linked to primary tumor formation, but also to metastases and cancer relapses. The knowledge that a tumor is organized hierarchically like normal tissues, namely comprising a small number of stem cells, which give rise to differentiated cells, thereby maintaining tissue integrity and organ function, is of crucial interest for our understanding how to treat cancer in future times [79]. The dilemma of current cancer therapies (conventional chemotherapy, radiation therapy, hormonal therapy, humanized monoclonal antibodies, and/or inhibitors) is that although most cancer patients respond to therapy, only a few are definitely cured [80], a matter which applies to both solid tumors as well as hematological disorders. This phenomenon, which has been entitled as 'the paradox of response and survival in cancer therapeutics' [80], has been compared to 'cutting a dandelion off at ground level' [80, 81]. Current cancer therapies are designed to target highly proliferating tumor cells and determination of tumor shrinking concomitant with mean disease-free survival of patients are commonly used as read-outs for the efficacy of the appropriate therapy. While such strategies eliminate the visible portion of the tumor, namely the tumor mass, they mostly fail to eliminate the unseen root of cancer, namely CSCs. Thus, elimination of the unseen root of cancer, CSCs, would mean to have a chance to cure the disease.

Because of their tumor-initiating capacity, CSCs have also been linked with metastasis formation and recurrences. In the context of metastasis formation, Li et al. [10] postulated the existence of a distinct type of CSCs exhibiting metastatic properties, the so-called mCSCs. We have recently postulated the existence of recurrence CSCs [82], which describes the type of CSCs that reinitiate tumor growth after first-line cancer therapy, thereby exhibiting an oncogenic resistance phenotype. Oncogenic resistance is associated with a highly aggressive cancer phenotype indicated by both an increased malignancy and drug resistance against first-line therapy [81].

If we agree that metastasis formation is initiated by circulating mCSCs, then we have to conclude that mCSCs must get accession to the circulation and must be capable of performing endothelial cell adhesion and transendothelial migration likely indicating that mCSCs and non-CSCs should use the same molecules for the rate-limiting step of extravasation [29]. Moreover, most cancers metastasize in an organ-specific manner, e.g. breast cancers preferentially metastasize into the regional lymph nodes, bone marrow, lung, and liver [53], whereas liver and lung are the preferred organs for metastasizing colon cancer cells [83]. Within the past years it became evident that the organ-specific metastatic spreading of tumor cells does not only rely on heterotypic and homotypic adhesive interactions, but also on the interplay of chemokines and their receptors [29]. For instance, breast cancer metastasis to lung and bone has been associated with $\alpha_v\beta_3$-integrin as well as CXCR4 and CCR7 expression [29]. Likewise, colon/colorectal cancer spreading to the liver is mediated by the selectins sialyl Lewis^a and sialyl Lewis^x, the integrins $\alpha_v\beta_3$ and $\alpha_v\beta_5$, as well as the chemokine receptors CXCR4 and CCR7 [29]. Thus, to induce metastases in an organ-specific manner, circulating mCSCs have to express some or all of the above-mentioned molecules.

Miki et al. [84] demonstrated that hTERT-immortalized malignant RC-92a tumor-derived prostate epithelial cells, which retained stem cell properties, responded to SDF-1α stimulation with an increased locomotory activity that was blocked using an anti-CXCR4 antibody. Recently, Hermann et al. [85] identified a distinct CSC subpopulation within pancreatic cancer that exhibited metastatic properties. Compared to CD133+ pancreatic CSCs, which were exclusively tumorigenic as well as highly resistant to standard chemotherapy, the metastatic pancreatic CSC variant also expressed the SDF-1α receptor CXCR4 [85]. Pancreatic mCSCs were identified in the invasive front of pancreatic tumors and only these cells determine the metastatic phenotype of the individual tumor. Depletion of pancreatic mCSCs from the total pancreatic CSC pool virtually abrogated metastasis formation of pancreatic tumors [85]. Similar results were obtained with the CXCR4 antagonist AMD3100, which significantly reduced pancreatic tumor metastasis in an animal model [85]. AMD3100 appears to be not only a potent agent for HSC/HSPC mobilization (see above), but may be also used as a compound that may help to reduce metastasis formation of various cancers due to the interruption of the SDF-1α/CXCR4 axis [86–93].

How mCSCs originate is not yet clear. One possibility could be that mCSC originate from primary tumor CSCs due to genetic instability. By contrast, Hermann et al. [85] provided evidence that the identified CD133+ CXCR4+ pancreatic mCSC represented a distinct invasive CSC population, which did not derive from original pancreatic tumor CD133+ CXCR4– CSCs, suggesting that mCSCs may originate independently from primary tumor CSCs. Wright et al. [94] reported recently that BRCA1 breast tumors contain distinct CD44+/CD24– and CD133+ cells with CSC characteristics. Thereby, cell lines derived from one tumor included increased numbers of CD44+/CD24– cells, which were previously identified as human breast CSCs, whereas cell lines derived from another mammary tumor exhibited low levels of CD44+/CD24– cells, but they harbored 2–5.9% CD133+ cells [94]. CD133 is not only a marker molecule for primitive HSCs/HSPCs [11], but also for some CSCs including brain CSCs [95] and colon CSCs [96]. These data show that one tumor can harbor distinct types of primary tumor CSCs, but it remains unclear whether these cells originated from a common precursor or independently from each other.

Hüsemann et al. [97] showed recently that the systemic spread is an early step in breast cancer. Hemizygous BALB-NeuT mice developed invasive mammary cancers within 23–30 weeks, whereby epithelial hyperplasia could already be detected microscopically in the mammary glands after 7–9 weeks [97, 98]. Progression to in situ carcinomas occurred between weeks 14 and 18, and at the same time tumors of the mammary gland became palpable or visible [97]. Investigation for cytokeratin (CK) and HER2 double positive breast cancer cells revealed that these cells became detectable in bone marrow at as early as 4–9 weeks when the most meticulous analysis of the mammary gland could detect areas of atypical ductal hyperplasia [97]. Likewise, single HER2-positive mammary tumor cells became detectable in lung tissue from week 9 on, and micrometastases were first visible around week 20 [97]. Resection of mammary glands of BALB-NeuT mice at week 18 neither prevented nor reduced the number of lung metastases, clearly indicating that dissemination of metastatic cancer cells had already occurred. Thus it can be assumed that the origin of mCSCs should also be an early event in cancer, whereby the way how these cells originates needs to be resolved in future work. This applies as well to the characterization of mCSCs and whether they are phenotypically similar of different to primary tumor CSCs.

As mentioned above, CSCs have come into the focus of cancer research. Specific CSC elimination strategies, e.g. by turning these cells from an inactive into an active state of the cell cycle, thereby making them susceptible for conventional cancer therapy [79], would make it possible to definitely cure cancer. However, the findings of Hermann et al. [85] indicate that primary tumor CSCs and mCSCs are two distinct CSC populations, which poses the question whether both CSC populations could be treated with one common anti-CSC strategy or whether CSC subpopulation-specific strategies have to be developed. Even if it would not be possible to eliminate CSCs specifically, the success of such strategies depends on molecules being exclusively expressed by CSCs and not by normal stem cells. The necessity of mCSCs to migrate might be a useful target for appropriate therapeutical approaches. Although inhibition of cell migration will not eliminate CSCs, it may help to delay formation of metastases, which is still the primary cause of death in cancer.

Conclusion

The ability to migrate is a prerequisite for various stem cell types in order to facilitate their biological functions within the body. For HSCs/HSPCs the ability to migrate is mandatory for both mobilization and homing, whereby the term 'homing' belongs to both bone marrow repopulation as well as recruitment of HSCs/HSPCs into degenerated tissues. An impairment of homing processes, e.g. HSCs/HSPCs do not respond to the chemoattractant stimuli SDF-1α or fail to engraft, would have fatal consequences for individuals who have received a bone marrow stem cell transplantation. We have recently demonstrated that culture conditions do influence the migratory activity of murine Lin$^-$ c-kit$^+$ HSPCs [1]. Thereby, cultivation of murine Lin$^-$ c-kit$^+$ HSPCs in the presence of [Flt3-ligand, IL-11] gave rise to cells which do exhibit a functional SDF-1α/CXCR4 axis (as indicated by induction of CXCR4-specific signal transduction cascades), but which do not respond to SDF-1α with an increased locomotory activity (fig. 1) [1]. Preliminary in vivo homing data (cultivated murine Lin$^-$ c-kit$^+$ HSPCs were labeled with CSFE, injected into the tail vein of recipient mice and distribution of labeled cells in the peripheral blood, spleen and bone marrow was determined 18 h later by flow cytometry) revealed that the amount of cultivated murine Lin$^-$ c-kit$^+$ HSPCs, which showed a weak SDF-1α-mediated migratory activity, within the bone marrow was markedly decreased or even not detectable as compared to cultivated cells showing a high SDF-1α-mediated migratory activity [S. Kassmer and T. Dittmar, unpubl. results]. These findings nicely illustrate the necessity of HSCs/HSPCs to migrate in order to fulfill their biological function.

The role of the modulation of the SDF-1α-induced migration of HSCs/HSPCs by e.g. IL-8 or GABA, and whether such processes are involved during inflammatory conditions needs to be clarified in future studies. This applies as well to considerations whether such mechanisms could be used for regenerative purposes, e.g. to ensure that stem cells reside at the appropriate place where they have been administered. In this context, the $GABA_B$-receptor agonist baclofen could be used for such purposes. Baclofen is a well-known drug, being already developed in the 1920s, which is used in human medicine to treat, e.g., spasticity of multiple sclerosis patients. However, as mentioned above, local application of stem cells into a degenerated organ tissue prerequisites that the administered cells are still migratory active in order to regenerate a three-dimensional organ environment. Thus appropriate studies have to be performed first to investigate the regenerative capacity of HSCs/HSPCs in the presence of compounds that modulate the SDF-1α/CXCR4 axis.

In case of CSC migration, the modulation of specific promigratory pathways might be useful to impair/slow down metastasis formation. As mentioned above, mCSCs may metastasize via the SDF-1α/CXCR4 axis, which can be used as a target. Impairment of CXCR4 signaling by AMD3100 significantly reduced metastasis formation of circulating CD133+CXCR4− pancreatic mCSCs in an animal model [85]. In addition to pancreatic cancer [85, 86, 89], AMD3100 also reduced metastasis formation of breast cancer [92] and human malignant melanoma [90]. Thus, the SDF-1α/CXCR4 axis might be an appropriate target to delay or even inhibit mCSC-mediated metastases formation. However, since inhibition of the SDF-1α/CXCR4 axis also affects HSCs/HSPCs, the use of AMD3100 in cancer treatment needs further investigations. As mentioned above, AMD3100 has recently been approved by the FDA as a HSC/HSPC-mobilizing drug. Inhibition of the SDF-1α/CXCR4 axis may also impair HSCs/HSPCs homing and may also affect the SDF-1α-mediated migration of other cells, e.g. lymphocytes.

In summary, the ability to migrate is a prerequisite for various stem cells to fulfill their biological function. In the context of HSCs/HSPCs and mCSCs, the modulation of the cells' migration might be used for the optimization of stem cell-based regeneration strategies or to delay metastasis formation.

Acknowledgements

Financial support by the Verein zur Förderung der Krebsforschung e.V., Heidelberg, Germany, and the Fritz-Bender-Foundation, Munich, Germany.

References

1 Kassmer SH, Niggemann B, Punzel M, Mieck C, Zanker KS, Dittmar T: Cytokine combinations differentially influence the SDF-1α-dependent migratory activity of cultivated murine hematopoietic stem and progenitor cells. Biol Chem 2008;389:863–872.

2 Aiuti A, Webb IJ, Bleul C, Springer T, Gutierrez-Ramos JC: The chemokine SDF-1 is a chemoattractant for human CD34+ hematopoietic progenitor cells and provides a new mechanism to explain the mobilization of CD34+ progenitors to peripheral blood. J Exp Med 1997;185:111–120.

3 Kasenda B, Kassmer SH, Niggemann B, Schiermeier S, Hatzmann W, Zanker KS, Dittmar T: The stromal cell-derived factor-1α-dependent migration of human cord blood CD34 haematopoietic stem and progenitor cells switches from protein kinase C (PKC)-α dependence to PKC-α independence upon prolonged culture in the presence of Flt3-ligand and interleukin-6. Br J Haematol 2008;142:831–835.

4 Weidt C, Niggemann B, Hatzmann W, Zanker KS, Dittmar T: Differential effects of culture conditions on the migration pattern of stromal cell-derived factor-stimulated hematopoietic stem cells. Stem Cells 2004;22:890–896.

5 Kim JB, Zaehres H, Wu G, Gentile L, Ko K, Sebastiano V, Arauzo-Bravo MJ, Ruau D, Han DW, Zenke M, Scholer HR: Pluripotent stem cells induced from adult neural stem cells by reprogramming with two factors. Nature 2008;454:646–650.

6 Takahashi K, Tanabe K, Ohnuki M, Narita M, Ichisaka T, Tomoda K, Yamanaka S: Induction of pluripotent stem cells from adult human fibroblasts by defined factors. Cell 2007;131:861–872.

7 Zhou H, Wu S, Joo JY, Zhu S, Han DW, Lin T, Trauger S, Bien G, Yao S, Zhu Y, Siuzdak G, Scholer HR, Duan L, Ding S: Generation of induced pluripotent stem cells using recombinant proteins. Cell Stem Cell 2009;4:381–384.

8 Clarke MF, Dick JE, Dirks PB, Eaves CJ, Jamieson CH, Jones DL, Visvader J, Weissman IL, Wahl GM: Cancer stem cells – perspectives on current status and future directions: AACR Workshop on Cancer Stem Cells. Cancer Res 2006;66:9339–9344.

9 Wicha MS, Liu S, Dontu G: Cancer stem cells: an old idea – a paradigm shift. Cancer Res 2006;66:1883–1896.

10 Li F, Tiede B, Massague J, Kang Y: Beyond tumorigenesis: cancer stem cells in metastasis. Cell Res 2007;17:3–14.

11 De Wynter EA, Buck D, Hart C, Heywood R, Coutinho LH, Clayton A, Rafferty JA, Burt D, Guenechea G, Bueren JA, Gagen D, Fairbairn LJ, Lord BI, Testa NG: CD34+AC133+ cells isolated from cord blood are highly enriched in long-term culture-initiating cells, NOD/SCID-repopulating cells and dendritic cell progenitors. Stem Cells 1998; 16:387–396.

12 Cottler-Fox MH, Lapidot T, Petit I, Kollet O, DiPersio JF, Link D, Devine S: Stem cell mobilization. Hematology Am Soc Hematol Educ Program 2003;419–437.

13 Weidt C, Niggemann B, Kasenda B, Drell TL, Zänker KS, Dittmar T: Stem cell migration: a quintessential stepping stone to successful therapy. Curr Stem Cell Res Treat 2007;2:89–103.

14 Levesque JP, Hendy J, Takamatsu Y, Williams B, Winkler IG, Simmons PJ: Mobilization by either cyclophosphamide or granulocyte colony-stimulating factor transforms the bone marrow into a highly proteolytic environment. Exp Hematol 2002;30:440–449.

15 Pelus LM, Bian H, King AG, Fukuda S: Neutrophil-derived MMP-9 mediates synergistic mobilization of hematopoietic stem and progenitor cells by the combination of G-CSF and the chemokines GROβ/CXCL2 and GROβT/CXCL2δ4. Blood 2004;103: 110–119.

16 Pruijt JF, Fibbe WE, Laterveer L, Pieters RA, Lindley IJ, Paemen L, Masure S, Willemze R, Opdenakker G: Prevention of interleukin-8-induced mobilization of hematopoietic progenitor cells in rhesus monkeys by inhibitory antibodies against the metalloproteinase gelatinase B (MMP-9). Proc Natl Acad Sci USA 1999;96:10863–10868.

17 Levesque JP, Takamatsu Y, Nilsson SK, Haylock DN, Simmons PJ: Vascular cell adhesion molecule-1 (CD106) is cleaved by neutrophil proteases in the bone marrow following hematopoietic progenitor cell mobilization by granulocyte colony-stimulating factor. Blood 2001;98:1289–1297.

18 Bonig H, Wundes A, Chang KH, Lucas S, Papayannopoulou T: Increased numbers of circulating hematopoietic stem/progenitor cells are chronically maintained in patients treated with the CD49d blocking antibody natalizumab. Blood 2008;111: 3439–3441.

19 Zohren F, Toutzaris D, Klarner V, Hartung HP, Kieseier B, Haas R: The monoclonal anti-VLA-4 antibody natalizumab mobilizes CD34+ hematopoietic progenitor cells in humans. Blood 2008;111 3893–3895.

20 Levesque JP, Hendy J, Takamatsu Y, Simmons PJ, Bendall LJ: Disruption of the CXCR4/CXCL12 chemotactic interaction during hematopoietic stem cell mobilization induced by GCSF or cyclophosphamide. J Clin Invest 2003;111:187–196.

21 Christopherson KW, Hangoc G, Broxmeyer HE: Cell surface peptidase CD26/dipeptidylpeptidase IV regulates CXCL12/stromal cell-derived factor-1α-mediated chemotaxis of human cord blood CD34+ progenitor cells. J Immunol 2002;169:7000–7008.

22 Christopherson KW 2nd, Cooper S, Broxmeyer HE: Cell surface peptidase CD26/DPPIV mediates G-CSF mobilization of mouse progenitor cells. Blood 2003;101:4680–4686.

23 Christopherson KW 2nd, Uralil SE, Porecha NK, Zabriskie RC, Kidd SM, Ramin SM: G-CSF- and GM-CSF-induced upregulation of CD26 peptidase downregulates the functional chemotactic response of CD34+CD38– human cord blood hematopoietic cells. Exp Hematol 2006;34:1060–1068.

24 Uy GL, Rettig MP, Cashen AF: Plerixafor, a CXCR4 antagonist for the mobilization of hematopoietic stem cells. Expert Opin Biol Ther 2008;8:1797–1804.

25 Broxmeyer HE, Orschell CM, Clapp DW, Hangoc G, Cooper S, Plett PA, Liles WC, Li X, Graham-Evans B, Campbell TB, Calandra G, Bridger G, Dale DC, Srour EF: Rapid mobilization of murine and human hematopoietic stem and progenitor cells with AMD3100, a CXCR4 antagonist. J Exp Med 2005;201:1307–1318.

26 Lapidot T: Mechanism of human stem cell migration and repopulation of NOD/SCID and $B2m^{null}$ NOD/SCID mice. The role of SDF-1/CXCR4 interactions. Ann NY Acad Sci 2001;938:83–95.

27 Wang X, Willenbring H, Akkari Y, Torimaru Y, Foster M, Al-Dhalimy M, Lagasse E, Finegold M, Olson S, Grompe M: Cell fusion is the principal source of bone-marrow-derived hepatocytes. Nature 2003;422:897–901.

28 Madri JA, Graesser D: Cell migration in the immune system: the evolving inter-related roles of adhesion molecules and proteinases. Dev Immunol 2000;7: 103–116.

29 Dittmar T, Heyder C, Gloria-Maercker E, Hatzmann W, Zanker KS: Adhesion molecules and chemokines: the navigation system for circulating tumor (stem) cells to metastasize in an organ-specific manner. Clin Exp Metastasis 2008;25:11–32.

30 Naiyer AJ, Jo DY, Ahn J, Mohle R, Peichev M, Lam G, Silverstein RL, Moore MA, Rafii S: Stromal derived factor-1-induced chemokinesis of cord blood CD34+ cells (long-term culture-initiating cells) through endothelial cells is mediated by E-selectin. Blood 1999;94:4011–4019.

31 Spertini O, Cordey AS, Monai N, Giuffre L, Schapira M: P-selectin glycoprotein ligand 1 is a ligand for L-selectin on neutrophils, monocytes, and CD34+ hematopoietic progenitor cells. J Cell Biol 1996;135:523–531.

32 Hidalgo A, Sanz-Rodriguez F, Rodriguez-Fernandez JL, Albella B, Blaya C, Wright N, Cabanas C, Prosper F, Gutierrez-Ramos JC, Teixido J: Chemokine stromal cell-derived factor-1α modulates VLA-4 integrin-dependent adhesion to fibronectin and VCAM-1 on bone marrow hematopoietic progenitor cells. Exp Hematol 2001;29:345–355.

33 Peled A, Kollet O, Ponomaryov T, Petit I, Franitza S, Grabovsky V, Slav MM, Nagler A, Lider O, Alon R, Zipori D, Lapidot T: The chemokine SDF-1 activates the integrins LFA-1, VLA-4, and VLA-5 on immature human CD34+ cells: role in transendothelial/stromal migration and engraftment of NOD/SCID mice. Blood 2000;95:3289–3296.

34 Solanilla A, Grosset C, Duchez P, Legembre P, Pitard V, Dupouy M, Belloc F, Viallard JF, Reiffers J, Boiron JM, Coulombel L, Ripoche J: Flt3-ligand induces adhesion of haematopoietic progenitor cells via a very late antigen (VLA)-4- and VLA-5-dependent mechanism. Br J Haematol 2003;120:782–786.

35 Yong KL, Watts M, Shaun TN, Sullivan A, Ings S, Linch DC: Transmigration of CD34+ cells across specialized and nonspecialized endothelium requires prior activation by growth factors and is mediated by PECAM-1 (CD31). Blood 1998;91:1196–1205.

36 Lapidot T, Kollet O: The essential roles of the chemokine SDF-1 and its receptor CXCR4 in human stem cell homing and repopulation of transplanted immune-deficient NOD/SCID and NOD/SCID/B2m^{null} mice. Leukemia 2002;16:1992–2003.

37 Peled A, Petit I, Kollet O, Magid M, Ponomaryov T, Byk T, Nagler A, Ben-Hur H, Many A, Shultz L, Lider O, Alon R, Zipori D, Lapidot T: Dependence of human stem cell engraftment and repopulation of NOD/SCID mice on CXCR4. Science 1999;283:845–848.

38 Peled A, Grabovsky V, Habler L, Sandbank J, Arenzana-Seisdedos F, Petit I, Ben-Hur H, Lapidot T, Alon R: The chemokine SDF-1 stimulates integrin-mediated arrest of CD34+ cells on vascular endothelium under shear flow. J Clin Invest 1999;104:1199–1211.

39 Servant G, Weiner OD, Herzmark P, Balla T, Sedat JW, Bourne HR: Polarization of chemoattractant receptor signaling during neutrophil chemotaxis. Science 2000;287:1037–1040.

40 Van Buul JD, Voermans C, van Gelderen J, Anthony EC, van der Schoot CE, Hordijk PL: Leukocyte-endothelium interaction promotes SDF-1-dependent polarization of CXCR4. J Biol Chem 2003;278:30302–30310.

41 Kollet O, Shivtiel S, Chen YQ, Suriawinata J, Thung SN, Dabeva MD, Kahn J, Spiegel A, Dar A, Samira S, Goichberg P, Kalinkovich A, Arenzana-Seisdedos F, Nagler A, Hardan I, Revel M, Shafritz DA, Lapidot T: HGF, SDF-1, and MMP-9 are involved in stress-induced human CD34+ stem cell recruitment to the liver. J Clin Invest 2003;112:160–169.

42 Rosu-Myles M, Gallacher L, Murdoch B, Hess DA, Keeney M, Kelvin D, Dale L, Ferguson SS, Wu D, Fellows F, Bhatia M: The human hematopoietic stem cell compartment is heterogeneous for CXCR4 expression. Proc Natl Acad Sci USA 2000;97:14626–14631.

43 Lee Y, Gotoh A, Kwon HJ, You M, Kohli L, Mantel C, Cooper S, Hangoc G, Miyazawa K, Ohyashiki K, Broxmeyer HE: Enhancement of intracellular signaling associated with hematopoietic progenitor cell survival in response to SDF-1/CXCL12 in synergy with other cytokines. Blood 2002;99:4307–4317.

44 Chang C, Werb Z: The many faces of metalloproteases: cell growth, invasion, angiogenesis and metastasis. Trends Cell Biol 2001;11:S37–43.

45 Zheng Y, Sun A, Han ZC: Stem cell factor improves SCID-repopulating activity of human umbilical cord blood-derived hematopoietic stem/progenitor cells in xenotransplanted NOD/SCID mouse model. Bone Marrow Transplant 2005;35:137–142.

46 Zheng Y, Watanabe N, Nagamura-Inoue T, Igura K, Nagayama H, Tojo A, Tanosaki R, Takaue Y, Okamoto S, Takahashi TA: Ex vivo manipulation of umbilical cord blood-derived hematopoietic stem/progenitor cells with recombinant human stem cell factor can upregulate levels of homing-essential molecules to increase their transmigratory potential. Exp Hematol 2003;31:1237–1246.

47 Gonzalo JA, Lloyd CM, Peled A, Delaney T, Coyle AJ, Gutierrez-Ramos JC: Critical involvement of the chemotactic axis CXCR4/stromal cell-derived factor-1α in the inflammatory component of allergic airway disease. J Immunol 2000;165:499–508.

48 Buckley CD, Amft N, Bradfield PF, Pilling D, Ross E, Arenzana-Seisdedos F, Amara A, Curnow SJ, Lord JM, Scheel-Toellner D, Salmon M: Persistent induction of the chemokine receptor CXCR4 by TGF-β_1 on synovial T cells contributes to their accumulation within the rheumatoid synovium. J Immunol 2000;165:3423–3429.

49 Nanki T, Hayashida K, El-Gabalawy HS, Suson S, Shi K, Girschick HJ, Yavuz S, Lipsky PE: Stromal cell-derived factor-1-CXC chemokine receptor 4 interactions play a central role in CD4+ T cell accumulation in rheumatoid arthritis synovium. J Immunol 2000;165:6590–6598.
50 Ceradini DJ, Gurtner GC: Homing to hypoxia: HIF-1 as a mediator of progenitor cell recruitment to injured tissue. Trends Cardiovasc Med 2005;15: 57–63.
51 Lapidot T, Dar A, Kollet O: How do stem cells find their way home? Blood 2005;106:1901–1910.
52 Kucia M, Jankowski K, Reca R, Wysoczynski M, Bandura L, Allendorf DJ, Zhang J, Ratajczak J, Ratajczak MZ: CXCR4-SDF-1 signalling, locomotion, chemotaxis and adhesion. J Mol Histol 2004;35: 233–245.
53 Muller A, Homey B, Soto H, Ge N, Catron D, Buchanan ME, McClanahan T, Murphy E, Yuan W, Wagner SN, Barrera JL, Mohar A, Verastegui E, Zlotnik A: Involvement of chemokine receptors in breast cancer metastasis. Nature 2001;410:50–56.
54 Zlotnik A: Chemokines and cancer. Int J Cancer 2006;119:2026–2029.
55 Heyder C, Gloria-Maercker E, Hatzmann W, Niggemann B, Zanker KS, Dittmar T: Role of the β_1-integrin subunit in the adhesion, extravasation and migration of T24 human bladder carcinoma cells. Clin Exp Metastasis 2005;22:99–106.
56 Bleul CC, Fuhlbrigge RC, Casasnovas JM, Aiuti A, Springer TA: A highly efficacious lymphocyte chemoattractant, stromal cell-derived factor 1. J Exp Med 1996;184:1101–1109.
57 Kim CH, Broxmeyer HE: In vitro behavior of hematopoietic progenitor cells under the influence of chemoattractants: stromal cell-derived factor-1, steel factor, and the bone marrow environment. Blood 1998;91:100–110.
58 Ganju RK, Brubaker SA, Meyer J, Dutt P, Yang Y, Qin S, Newman W, Groopman JE: The α-chemokine, stromal cell-derived factor-1α, binds to the transmembrane G-protein-coupled CXCR-4 receptor and activates multiple signal transduction pathways. J Biol Chem 1998;273:23169–23175.
59 Wang JF, Park IW, Groopman JE: Stromal cell-derived factor-1α stimulates tyrosine phosphorylation of multiple focal adhesion proteins and induces migration of hematopoietic progenitor cells: roles of phosphoinositide-3 kinase and protein kinase C. Blood 2000;95:2505–2513.
60 Vila-Coro AJ, Rodriguez-Frade JM, Martin De Ana A, Moreno-Ortiz MC, Martinez AC, Mellado M: The chemokine SDF-1α triggers CXCR4 receptor dimerization and activates the JAK/STAT pathway. FASEB J 1999;13:1699–1710.
61 Clark EA, Brugge JS: Integrins and signal transduction pathways: the road taken. Science 1995;268:233–239.
62 Dittmar T, Brandt BH, Lang K, Zänker KS, Entschladen F: Lessons from tumor and immunocompetent cells. The quantitative engagement of ligand-receptor interactions modulates stop-and-go behavior as well as proliferation. Medicina (B Aires) 2000;60 Suppl 2:27–33.
63 Dittmar T, Hüsemann A, Schewe Y, Nofer JR, Niggemann B, Zänker KS, Brandt BH: Induction of cancer cell migration by epidermal growth factor is initiated by specific phosphorylation of tyrosine 1248 of c-erbB-2 receptor via EGFR. FASEB J 2002; 16:1823–1825.
64 Entschladen F, Zänker KS: Locomotion of tumor cells: a molecular comparison to migrating pre- and postmitotic leukocytes. J Cancer Res Clin Oncol 2000;126:671–681.
65 Fukuda S, Broxmeyer HE, Pelus LM: Flt3 ligand and the Flt3 receptor regulate hematopoietic cell migration by modulating the SDF-1α(CXCL12)/CXCR4 axis. Blood 2005;105:3117–3126.
66 Petit I, Goichberg P, Spiegel A, Peled A, Brodie C, Seger R, Nagler A, Alon R, Lapidot T: Atypical PKC-ζ regulates SDF-1-mediated migration and development of human CD34+ progenitor cells. J Clin Invest 2005;115:168–176.
67 Mochly-Rosen D: Localization of protein kinases by anchoring proteins: a theme in signal transduction. Science 1995;268:247–251.
68 Hofmann J: The potential for isoenzyme-selective modulation of protein kinase C. FASEB J 1997; 11:649–669.
69 Martiny-Baron G, Kazanietz MG, Mischak H, Blumberg PM, Kochs G, Hug H, Marme D, Schachtele C: Selective inhibition of protein kinase C isozymes by the indolocarbazole Gö6976. J Biol Chem 1993;268:9194–9197.
70 Seidel J, Niggemann B, Punzel M, Fischer J, Zänker KS, Dittmar T: The neurotransmitter γ-aminobutyric-acid is a potent inhibitor of the stromal cell-derived factor-1. Stem Cells Dev 2007;16:827–836.
71 Entschladen F, Drell TL, Lang K, Joseph J, Zaenker KS: Tumour-cell migration, invasion, and metastasis: navigation by neurotransmitters. Lancet Oncol 2004;5:254–258.
72 Lang K, Hatt H, Niggemann B, Zaenker KS, Entschladen F: A novel function for chemokines: downregulation of neutrophil migration. Scand J Immunol 2003;57:350–361.
73 Liesveld JL, Rosell K, Panoskaltsis N, Belanger T, Harbol A, Abboud CN: Response of human CD34+ cells to CXC, CC, and CX3C chemokines: implications for cell migration and activation. J Hematother Stem Cell Res 2001;10:643–655.

74 Rane MJ, Gozal D, Butt W, Gozal E, Pierce WM Jr, Guo SZ, Wu R, Goldbart AD, Thongboonkerd V, McLeish KR, Klein JB: γ-Aminobutyric acid type B receptors stimulate neutrophil chemotaxis during ischemia-reperfusion. J Immunol 2005;174:7242–7249.
75 Steidl U, Bork S, Schaub S, Selbach O, Seres J, Aivado M, Schroeder T, Rohr UP, Fenk R, Kliszewski S, Maercker C, Neubert P, Bornstein SR, Haas HL, Kobbe G, Tenen DG, Haas R, Kronenwett R: Primary human CD34+ hematopoietic stem and progenitor cells express functionally active receptors of neuromediators. Blood 2004;104:81–88.
76 Eyler CE, Rich JN: Survival of the fittest: cancer stem cells in therapeutic resistance and angiogenesis. J Clin Oncol 2008;26:2839–2845.
77 Rich JN: Cancer stem cells in radiation resistance. Cancer Res 2007;67:8980–8984.
78 Shervington A, Lu C: Expression of multidrug resistance genes in normal and cancer stem cells. Cancer Invest 2008;26:535–542.
79 Dittmar T, Zänker KS: Targeting cancer stem cells; in Dittmar T, Zänker KS (eds): Cancer and Stem Cells. Hauppauge, Nova Publishers, 2008, pp 189–197.
80 Huff CA, Matsui W, Smith BD, Jones RJ: The paradox of response and survival in cancer therapeutics. Blood 2006;107:431–434.
81 Blagosklonny MV: Why therapeutic response may not prolong the life of a cancer patient: selection for oncogenic resistance. Cell Cycle 2005;4:1693–1698.
82 Dittmar T, Nagler C, Schwitalla S, Reith G, Niggemann B, Zanker KS: Recurrence cancer stem cells – made by cell fusion? Med Hypotheses 2009;73:542–547.
83 Schluter K, Gassmann P, Enns A, Korb T, Hemping-Bovenkerk A, Holzen J, Haier J: Organ-specific metastatic tumor cell adhesion and extravasation of colon carcinoma cells with different metastatic potential. Am J Pathol 2006;169:1064–1073.
84 Miki J, Furusato B, Li H, Gu Y, Takahashi H, Egawa S, Sesterhenn IA, McLeod DG, Srivastava S, Rhim JS: Identification of putative stem cell markers, CD133 and CXCR4, in hTERT-immortalized primary nonmalignant and malignant tumor-derived human prostate epithelial cell lines and in prostate cancer specimens. Cancer Res 2007;67:3153–3161.
85 Hermann PC, Huber SL, Herrler T, Aicher A, Ellwart JW, Guba M, Bruns CJ, Heeschen C: Distinct populations of cancer stem cells determine tumor growth and metastatic activity in human pancreatic cancer. Cell Stem Cell 2007;1:313–323.
86 Marchesi F, Monti P, Leone BE, Zerbi A, Vecchi A, Piemonti L, Mantovani A, Allavena P: Increased survival, proliferation, and migration in metastatic human pancreatic tumor cells expressing functional CXCR4. Cancer Res 2004;64:8420–8427.
87 Ottaiano A, Franco R, Aiello Talamanca A, Liguori G, Tatangelo F, Delrio P, Nasti G, Barletta E, Facchini G, Daniele B, Di Blasi A, Napolitano M, Ierano C, Calemma R, Leonardi E, Albino V, De Angelis V, Falanga M, Boccia V, Capuozzo M, Parisi V, Botti G, Castello G, Vincenzo Iaffaioli R, Scala S: Overexpression of both CXC chemokine receptor 4 and vascular endothelial growth factor proteins predicts early distant relapse in stage II-III colorectal cancer patients. Clin Cancer Res 2006;12:2795–2803.
88 Rubin JB, Kung AL, Klein RS, Chan JA, Sun Y, Schmidt K, Kieran MW, Luster AD, Segal RA: A small-molecule antagonist of CXCR4 inhibits intracranial growth of primary brain tumors. Proc Natl Acad Sci USA 2003;100:13513–13518.
89 Saur D, Seidler B, Schneider G, Algul H, Beck R, Senekowitsch-Schmidtke R, Schwaiger M, Schmid RM: CXCR4 expression increases liver and lung metastasis in a mouse model of pancreatic cancer. Gastroenterology 2005;129:1237–1250.
90 Scala S, Giuliano P, Ascierto PA, Ierano C, Franco R, Napolitano M, Ottaiano A, Lombardi ML, Luongo M, Simeone E, Castiglia D, Mauro F, De Michele I, Calemma R, Botti G, Caraco C, Nicoletti G, Satriano RA, Castello G: Human melanoma metastases express functional CXCR4. Clin Cancer Res 2006;12:2427–2433.
91 Scotton CJ, Wilson JL, Scott K, Stamp G, Wilbanks GD, Fricker S, Bridger G, Balkwill FR: Multiple actions of the chemokine CXCL12 on epithelial tumor cells in human ovarian cancer. Cancer Res 2002;62:5930–5938.
92 Smith MC, Luker KE, Garbow JR, Prior JL, Jackson E, Piwnica-Worms D, Luker GD: CXCR4 regulates growth of both primary and metastatic breast cancer. Cancer Res 2004;64:8604–8612.
93 Yasumoto K, Koizumi K, Kawashima A, Saitoh Y, Arita Y, Shinohara K, Minami T, Nakayama T, Sakurai H, Takahashi Y, Yoshie O, Saiki I: Role of the CXCL12/CXCR4 axis in peritoneal carcinomatosis of gastric cancer. Cancer Res 2006;66:2181–2187.
94 Wright MH, Calcagno AM, Salcido CD, Carlson MD, Ambudkar SV, Varticovski L: Brca1 breast tumors contain distinct CD44+/CD24− and CD133+ cells with cancer stem cell characteristics. Breast Cancer Res 2008;10:R10.

95 Singh SK, Clarke ID, Terasaki M, Bonn VE, Hawkins C, Squire J, Dirks PB: Identification of a cancer stem cell in human brain tumors. Cancer Res 2003;63:5821–5828.
96 Ricci-Vitiani L, Lombardi DG, Pilozzi E, Biffoni M, Todaro M, Peschle C, De Maria R: Identification and expansion of human colon-cancer-initiating cells. Nature 2007;445:111–115.
97 Hüsemann Y, Geigl JB, Schubert F, Musiani P, Meyer M, Burghart E, Forni G, Eils R, Fehm T, Riethmuller G, Klein CA: Systemic spread is an early step in breast cancer. Cancer Cell 2008;13:58–68.
98 Di Carlo E, Diodoro MG, Boggio K, Modesti A, Modesti M, Nanni P, Forni G, Musiani P: Analysis of mammary carcinoma onset and progression in HER-2/neu oncogene transgenic mice reveals a lobular origin. Lab Invest 1999;79:1261–1269.

Dr. Thomas Dittmar
Institute of Immunology, Witten/Herdecke University
Stockumer Strasse 10, DE–58448 Witten (Germany)
Tel. +49 2302 926165, Fax +49 2302 926158, E-Mail thomas.dittmar@uni-wh.de

Entschladen F, Zänker KS (eds): Cell Migration: Signalling and Mechanisms.
Transl Res Biomed. Basel, Karger, 2010, vol 2, pp 28–39

Leukocyte Motility and Human Disease

Kate Cooper · Paul Nuzzi · Anna Huttenlocher

Departments of Pediatrics and Medical Microbiology and Immunology, University of Wisconsin-Madison, Madison, Wisc., USA

Abstract

Neutrophils are key mediators of the innate immune response and are often the first responders to inflammatory stimuli including tissue wounding and infection. Recruitment of neutrophils to inflamed tissues is essential for the normal host immune responses to infection but can also contribute to the development of chronic inflammatory disease. This chapter focuses on the mechanisms of neutrophil-directed migration and how defects in neutrophil motility or trafficking contribute to the pathogenesis of immunodeficiency and chronic inflammatory disease.

Basic Steps of Cell Movement

Cell migration requires a regulated and dynamic interaction between the cell and its surrounding environment [1]. To migrate, cells respond to directional cues by extending a localized protrusion or pseudopod in the direction of cell movement. For cell translocation to occur, the leading edge of the cell stabilizes an adhesion, which generates the traction required for cell movement. Subsequently, the cell must detach adhesions at the rear to allow for directed motility. Cell migration may therefore be separated into distinct steps including membrane protrusion and pseudopod formation, generation of contractile force and rear release [2]. The classic three-step migration pattern describes the migration patterns of mesenchymal cells including fibroblasts. In contrast, the mechanisms that govern the movement of the more rapidly moving cells of the immune system, such as neutrophils, appear to be distinct. In contrast to fibroblasts, neutrophils demonstrate intrinsic polarization and efficiently coordinate adhesion formation at the cell front and rear release, thereby demonstrating gliding migration [3].

External Factors That Regulate Cell Migration

Cell migration involves the integration of external cues, including factors that can either promote or inhibit cell motility. Furthermore, the cell needs to prioritize competing signals including competing gradients of chemoattractants or repellents. The external cues that regulate cell migration are diverse and include chemokines, growth factors, cell-cell contacts and extracellular matrix environment [4]. Although the majority of cells in adult organisms are non-migratory, leukocytes exhibit spontaneous motility and navigate through diverse extracellular environments as a normal part of immune surveillance. This capacity for invasive migration in what would normally be non-permissive tissues is also a feature of metastatic tumor cells. In response to specific tissue perturbations, such as wounding, the extracellular environment becomes permissive and contains migration-promoting signals that stimulate the recruitment of both leukocytes and fibroblast cells. Although the nature of the signals that mediate cell motility and leukocyte recruitment at tissue wounds have remained elusive, recent evidence suggests that a first step in leukocyte recruitment to a wound is the generation of a gradient of hydrogen peroxide at the wound [5]. Substantial evidence also implicates a role for growth factors and chemokines/inflammatory mediators in the subsequent recruitment and retention of leukocytes in inflamed tissues.

In addition to the presence of migration-promoting cues in the environment, there are also cues that inhibit cell migration. An important migration-inhibiting signal is mediated by the extracellular matrix. For example, high densities of fibronectin may inhibit cell migration. In fact, many cell types exhibit a biphasic relationship between adhesion and migration rate, with optimum speed occurring at an intermediate cell-substratum adhesion. Previous studies have suggested that motility is impaired at high ligand density because of reduced cell detachment [6]. Alternatively, high ligand-density may also regulate intracellular signaling and cell polarization/protrusion to affect cell migration speed [7]. More specifically, high fibronectin density down-regulates signaling pathways via Rac and Cdc42, critical for cell protrusion and polarization, thereby inhibiting cell migration [8]. Critical factors that influence cell motility also include inflammatory mediators that can induce a stop signal and contribute to leukocyte retention in inflamed tissues. An example includes the inflammatory mediator TNF-α that induces a neutrophil stop signal through the regulation of p38 MAPK signaling [9]. Another mechanism of migration inhibition involves cell-cell interaction and activation through the T-cell receptor, which induces a stop signal in T cells mediating prolonged T-cell signaling and leukocyte retention in inflamed tissues [10].

Cell Signaling during Neutrophil-Directed Migration

Neutrophils effectively respond to competing environmental cues and migrate rapidly up shallow gradients of chemoattractants. The resting neutrophil is maintained in a rounded, non-adherent state and, in response to either a gradient or uniform concentration of chemoattractant, adopts a polarized morphology. Chemotaxis is achieved by two distinct processes: the actin-independent sensing of chemoattractant gradients and the subsequent actin-dependent cell polarization, with the generation of a leading-edge pseudopod. A hallmark of the polarized morphology is the asymmetric recruitment of signaling molecules [reviewed in 11] (fig. 1, 2). However, the mechanism by which this is achieved and how the responses may differ to uniform concentrations or gradients of different chemoattractants remains poorly understood. In fact, studies suggest that neutrophils display a hierarchical response to external cues and are able to prioritize external stimuli. For example, a previous study demonstrates that neutrophils display preferential responses to end-target chemoattractants (fMLP) as compared to intermediary chemoattractants (IL-8) by the activation of specific signaling pathways [12, 13].

Sensing the Gradient

Neutrophils sense most chemoattractants via G-protein coupled receptors (GPCR) that are evenly distributed on the cell surface in a gradient of chemoattractant [14]. Neutrophils are able to sense bacterial products, such as lipopolysaccride (LPS) or fMLP. Additionally, the cells have receptors that recognize chemicals produced by the host such as chemokines and cytokines, including the complement factor 5a (C5a) fragment.

The cell is able to accomplish directed cell motility by amplifying the external existing gradient of chemoattractant into a steeper internal gradient of cell signaling. One important molecule implicated in gradient sensing is the PI3K product phosphatidylinositol (3,4,5)-trisphosphate (PtdIns(3,4,5)P_3). When the cell is exposed to a gradient of chemoattractant, the highest concentration of (PtdIns(3,4,5)P_3) is along the membrane closest to the source of chemoattractant [15, 16]. There is evidence from the model organism *Dictyostelium* that PI3K localizes to the front of the cell and PTEN that converts PtdIns(3,4,5)P_3 into phosphatidylinositol (4,5)-bisphosphate (PtdIns(4,5)P_2) is located at the cell rear, together contributing to the segregation of phosphoinositide signaling during chemotaxis [17, 18]. The higher concentration of PtdIns(3,4,5)P_3 along the membrane at the leading provides binding sites for other proteins that are important for amplifying the asymmetry in cell signaling leading to actin polymerization at the cell front.

Rho GTPase Signaling during Chemotaxis

Substantial evidence implicates a critical role for Rho GTPase signaling during chemotaxis. Many of the activators of the small GTPases, guanine nucleotide

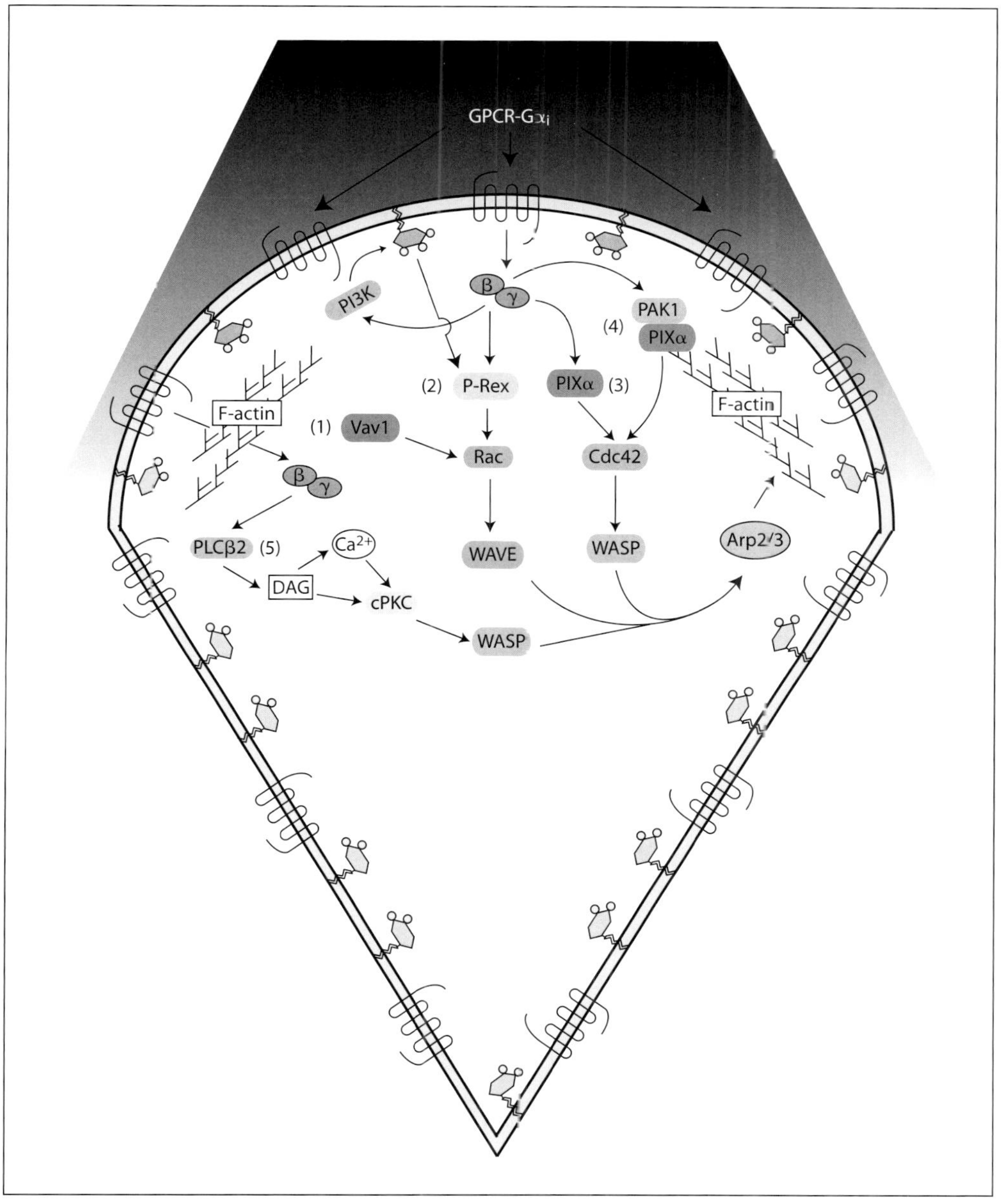

Fig. 1. Regulation of polarity at the leading edge. Schematic of polarized signaling events at the leading edge of neutrophils in a gradient of chemoattractant.

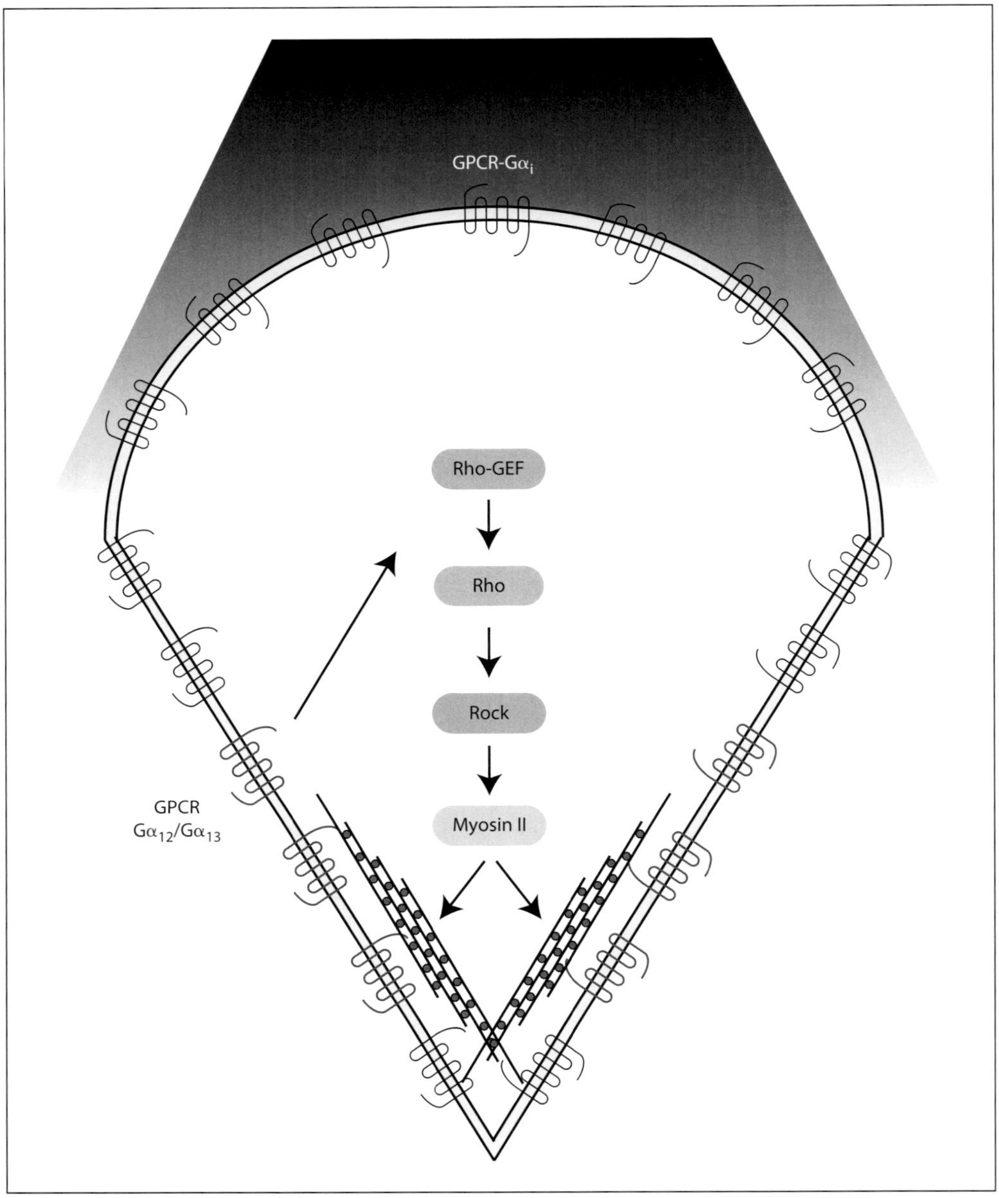

Fig. 2. Regulation of polarity at the uropod of neutrophils. Schematic of polarized signaling events at the neutrophil uropod during chemotaxis in a gradient of chemoattractant.

exchange factors (GEFs) have plextrin homology (PH) domains, which have been found to translocate to the side of the cell that is facing the highest concentration of chemoattractant [15, 16]. Indeed the RacGEF, P-Rex1, has been found to be activated by PtdIns(3,4,5)P_3 and localizes to the leading edge during chemotaxis [19], suggesting that in neutrophils Rac provides a key bridge between the gradient-sensing machinery and localized activation of actin polymerization at the cell front. While Rac is important at the leading edge [20], additional small GTPases are also involved in neutrophil chemotaxis. Cdc42 maintains stability and location of the pseudopod extensions [20], while also limiting pseudopod formation at the cell rear [21]. In contrast, RhoA is involved with mediating contractility and detachment of the cell rear [22].

Cytoskeleton and Neutrophil Motility

Asymmetric actin polymerization is critical for directed neutrophil motility. Therefore, many components of the actin regulatory machinery are localized to the leading edge of the neutrophil. Arp2/3 is an important nucleator of actin polymerization and binds to other regulatory proteins including Wiskott-Aldrich syndrome protein (WASP). WASP activates actin polymerization through its action on Arp2/3 and is activated by the small GTPase Cdc42 [23]. This signaling pathway therefore likely provides a mechanism by which the Cdc42 affects the stability and localization of pseudopod formation during directed cell migration. Additionally, other actin-binding and modifying proteins including Scar/WAVE proteins, and actin-severing proteins like cofilin have been implicated in neutrophil-directed migration [24].

Microtubules are also important regulators of neutrophil chemotaxis. Microtubules orient toward the uropod or cell rear of migrating neutrophils [25, 26]. Interestingly, microtubules are not necessary for neutrophil polarization but mediate neutrophil-directed migration since disruption of microtubules impairs neutrophil chemotaxis [25, 27, 28]. It has been hypothesized that the microtubules can affect cell motility by regulating the activity of small GTPases in specific cellular locations during cell migration [28].

Recent studies have examined the cytoskeletal force generated by neutrophils as they migrate. Interestingly, the force on the substratum is the highest underneath the rear of the migrating neutrophil, in contrast to fibroblasts [29]. Accordingly, myosin II has been localized toward the uropod of amoeboid cells including *Dictyostelium* and neutrophils. The rearward actomyosin contractility generates sufficient force to propel the cell forward, in a Rho GTPase-dependent fashion [28, 30]. This activation of actomyosin contractility is necessary for rear retraction [31, 32]. For example, inhibition of Rho or ROCK activity impairs cell detachment of migrating leukocytes [33].

Neutrophil Motility in Disease

Analysis of neutrophil function from patients with immune disorders can provide important information about neutrophil motility and chemotaxis. Defects in neutrophil polarization and directed migration have been observed in patients with immune deficiency disorders, including leukocyte adhesion deficiency and lazy leukocyte syndrome [reviewed in 34], and more recently in patients with chronic inflammatory disease [35].

Immunodeficiency Diseases

Leukocyte adhesion deficiency is a rare autosomal recessive disease characterized by recurrent bacterial infections and abscess formation [36]. The genetic defect is most commonly a mutation in the β_2 integrin (CD18) gene that impairs surface expression of β_2 integrins. Leukocyte transmigration and recruitment to tissues are severely impaired in patients with leukocyte adhesion deficiency [37], implicating an essential role for β_2 integrins in neutrophil function.

Signaling molecules downstream of integrins and receptor activation have also been implicated in immunodeficiency disorders. For example, Rac2 is a hematopoietic-specific family member that has been shown to be crucial for proper neutrophil chemotaxis into inflamed tissue [38]. A patient with a dominant mutation in Rac2, D57N, has been reported who presented with recurrent bacterial infections and impaired neutrophil polarization and chemotaxis [38], indicating an essential role for Rac2 in neutrophil-directed migration and highlighting the role of migration in immunodeficiencies.

Defects in chemokine receptor function have also been reported in patients with neutropenia and recurrent bacterial infection. Patients with WHIM syndrome (warts, hypogammaglobulinemia, recurrent bacteria infections, and myelokathexis) have neutrophil retention in the bone marrow and neutropenia resulting in recurrent bacterial infections. Patients with the dominantly-inherited WHIM syndrome have mutations in CXCR4, a receptor that recognizes CXCL12 (SDF-1) [39]. WHIM syndrome-associated mutations have a truncation of the cytoplasmic carboxyl-(C)-terminal tail of the receptor preventing downregulation of the receptor [40]. Accordingly, neutrophils from patients with WHIM syndrome show increased chemotaxis of neutrophils to CXCL12 [41]. CXCL12 is involved in the homing of senescent neutrophils to the bone marrow and retains immature neutrophils within the bone marrow [42]. Since mature neutrophils are not normally responsive to the chemokine [43], it is likely that increased signaling through the CXCR4 receptor may retain neutrophils in the bone marrow resulting in neutropenia and an increased susceptibility to bacterial infections. Although the exact molecular mechanism that contributes to the development of WHIM syndrome has not been determined, substantial evidence implicates abnormal signaling through CXCR4 in disease pathogenesis.

Wiskott-Aldrich syndrome (WAS) is another immunodeficiency associated with defects in leukocyte motility. WAS patient macrophages have been reported to have severely impaired chemotaxis [44, 45]. WAS is caused by mutations in WASP which is a hematopoietic-specific protein regulated by Cdc42 that activates Arp2/3 and leads to actin polymerization. Migration defects in leukocytes form patients with WAS are likely due to abnormal regulation of the actin cytoskeleton.

Neutrophil Motility in Chronic Inflammatory Disease

Autoinflammatory diseases are characterized by unprovoked inflammation and tissue destruction involving cells of the innate immune system. A variety of symptoms may be present with these diseases including periodic fever, urticaria-like rash, sterile arthritis, sensorineural hearing loss, sterile peritonitis, and inflammation of joints and skin [46]. While the autoinflammatory diseases are clinically well defined, less is known about the molecular mechanisms that contribute to disease pathogenesis. Many autoinflammatory diseases are hereditary and monogenetic including familial Mediterranean fever (FMF) and the cryopyrin-associated periodic syndromes (CAPS) including neonatal-onset multisystem inflammatory disease (NOMID), Muckle-Wells syndrome (MWS), and familial cold autoinflammatory syndrome (FCAS). Pyogenic sterile arthritis with pyoderma gangrenosum and acne (PAPA) syndrome and tumor necrosis factor receptor (TNFR)-associated periodic syndrome (TRAPS) are also inherited autoinflammatory diseases. Other common and more genetically complex autoinflammatory diseases such as Crohn's disease and gout can show similarities to these classic autoinflammatory diseases in symptoms and some available treatments [47]. The genes that are mutated in many of the hereditary autoinflammatory diseases have been determined, and there has been much recent work to further our understanding of the intracellular pathways affected by these mutations.

Importantly, many of these proteins are components of a proinflammatory multiprotein complex called the inflammasome. The intracellular inflammasome senses danger by recognizing the presence of pathogen-associated molecules, such as microbial motifs and toxins, live bacteria, and viruses, as well as danger-associated host components like ATP and monosodium urate crystals [reviewed in 48, 49]. Activation of the inflammasome mediates the conversion of prointerleukin (pro-IL)-1β to its active form (IL-1β). The NLRP3 inflammasome is the most well-characterized complex and importantly, NLRP3 (formerly referred to as NALP3 or CIAS1, which encodes cryopyrin [50]) is mutated in CAPS. Recently it was discovered that the NLRP3 inflammasome is also involved in mediating inflammation caused by viral and host cytosolic DNA [51]. The proteins pyrin and PSTPIP1, involved in FMF and PAPA syndrome respectively, have also been linked to the production of IL-1β [52–55]. Substantial evidence suggests that the dysregulated production and release of the inflammatory mediator IL-1β is an important factor that contributes to many of the disease manifestations, since patients clinically respond to IL-1β antagonists [49].

A hallmark of autoinflammatory diseases is abnormal neutrophil infiltration into tissues, suggesting defects in neutrophil trafficking [46]. For example, neutrophils accumulate in the arthritic joints of PAPA patients [56] and biopsies of inflamed serosal membranes, joints and skin of patients with FMF show primarily neutrophils [46]. Additionally, the uritcarial rash of NOMID/MWS is characterized by infiltration of neutrophils [57].

In contrast to primary immunodeficiency disorders, few studies have addressed the contribution of leukocyte motility to the pathogenesis of autoinflammatory disease. It is an intriguing possibility that defects in leukocyte motility and trafficking may also contribute to the pathogenesis of autoinflammatory diseases. In fact, many patients with autoinflammatory disease, including FMF, respond to agents that affect cell migration, including treatment with the microtubule-disrupting drug colchicine, suggesting that drugs that target neutrophil motility or trafficking may have beneficial effects in patients with autoinflammatory disorders [58, 59].

Recent studies suggest that patients with autoinflammatory disorders may have defects in neutrophil motility. For example, neutrophils from patients with mutations in cryopyrin (NALP3) have been reported to have impaired chemotaxis [35, 57]. These findings suggest that defects in neutrophil chemotaxis may contribute to the pathogenesis of NOMID/MWS. Furthermore, autoinflammatory disease-associated mutations have been reported in proteins with known roles in cell motility. For example, PAPA syndrome patients have mutations in the adaptor protein PSTPIP1 that binds to WASP [56], and regulates neutrophil motility [60]. Patients with TNFR-1-associated periodic syndrome (TRAPS) have mutations in the TNF receptor that enhance TNF receptor signaling [61]. Patients with TRAPS often benefit from treatment with agents that block TNF-α signaling including the soluble TNF p75 receptor antagonist, etanercept and anti-TNF-α monoclonal antibodies that both block TNF activity. TNF-α is a potent proinflammatory cytokine that has been reported to modulate neutrophil adhesion and migration, and contribute to neutrophil retention in inflamed tissue. Future studies will provide further insight into how defects in leukocyte adhesion or trafficking can contribute to the pathogenesis of chronic inflammatory disease.

Conclusion

There has been substantial progress in the last decade in understanding the mechanisms that regulate leukocyte motility and chemotaxis. Despite recent progress, our understanding of the mechanisms that regulate directional migration in complex in vivo environments and the integration of diverse external cues remains limited. A challenge for future investigation will be to understand how defects in neutrophil motility or trafficking contribute to the pathogenesis of both immune and inflammatory disorders and how these pathways can be targeted to treat human disease.

References

1 Ridley AJ, Schwartz MA, Burridge K, Firtel RA, Ginsberg MH, Borisy G, Parsons JT, Horwitz AR: Cell migration: integrating signals from front to back. Science 2003;302:1704–1709.

2 Stossel TP: On the crawling of animal cells. Science 1993;260:1086–1094.

3 Cox EA, Huttenlocher A: Regulation of integrin-mediated adhesion during cell migration. Microsc Res Tech 1998;43:412–419.

4 Lauffenburger DA: Cell motility. Making connections count. Nature 1996;383:390–391.

5 Niethammer P, Grabher C, Look AT, Mitchison TJ: A tissue-scale gradient of hydrogen peroxide mediates rapid wound detection in zebrafish. Nature 2009;459:996–999.

6 Huttenlocher A, Sandborg RR, Horwitz AF: Adhesion in cell migration. Curr Opin Cell Biol 1995;7:697–706.

7 Cox EA, Sastry SK, Huttenlocher A: Integrin-mediated adhesion regulates cell polarity and membrane protrusion through the Rho family of GTPases. Mol Biol Cell 2001;12:265–277.

8 Gupton SL, Collings DA, Allen NS: Endoplasmic reticulum targeted GFP reveals ER organization in tobacco NT-1 cells during cell division. Plant Physiol Biochem 2006;44:95–105.

9 Lokuta MA, Huttenlocher A: TNF-α promotes a stop signal that inhibits neutrophil polarization and migration via a p38 MAPK pathway. J Leukoc Biol 2005;78:210–219.

10 Rudd CE: The reverse stop-signal model for CTLA4 function. Nat Rev Immunol 2008;8:153–160.

11 Van Haastert PJ, Devreotes PN: Chemotaxis: signalling the way forward. Nat Rev Mol Cell Biol 2004;5:626–634.

12 Heit B, Tavener S, Raharjo E, Kubes P: An intracellular signaling hierarchy determines direction of migration in opposing chemotactic gradients. J Cell Biol 2002;159:91–102.

13 Heit B, Liu L, Colarusso P, Puri KD, Kubes P: PI3K accelerates, but is not required for, neutrophil chemotaxis to fMLP. J Cell Sci 2008;121:205–214.

14 Servant G, Weiner OD, Neptune ER, Sedat JW, Bourne HR: Dynamics of a chemoattractant receptor in living neutrophils during chemotaxis. Mol Biol Cell 1999;10:1163–1178.

15 Servant G, Weiner OD, Herzmark P, Balla T, Sedat JW, Bourne HR: Polarization of chemoattractant receptor signaling during neutrophil chemotaxis. Science 2000;287:1037–1040.

16 Parent CA, Blacklock BJ, Froehlich WM, Murphy DB, Devreotes PN: G-protein signaling events are activated at the leading edge of chemotactic cells. Cell 1998;95:81–91.

17 Iijima M, Devreotes P: Tumor suppressor PTEN mediates sensing of chemoattractant gradients. Cell 2002;109:599–610.

18 Funamoto S, Meili R, Lee S, Parry L, Firtel RA: Spatial and temporal regulation of 3-phosphoinositides by PI 3-kinase and PTEN mediates chemotaxis. Cell 2002;109:611–623.

19 Welch HC, Coadwell WJ, Ellson CD, Ferguson GJ, Andrews SR, Erdjument-Bromage H, Tempst P, Hawkins PT, Stephens LR: P-Rex1, a PtdIns(3,4,5)P_3- and Gβγ-regulated guanine-nucleotide exchange factor for Rac. Cell 2002;108:809–821.

20 Srinivasan S, Wang F, Glavas S, Ott A, Hofmann F, Aktories K, Kalman D, Bourne HR: Rac and Cdc42 play distinct roles in regulating PI(3,4,5)P_3 and polarity during neutrophil chemotaxis. J Cell Biol 2003;160:375–385.

21 Van Keymeulen A, Wong K, Knight ZA, Govaerts C, Hahn KM, Shokat KM, Bourne HR: To stabilize neutrophil polarity, PIP_3 and Cdc42 augment RhoA activity at the back as well as signals at the front. J Cell Biol 2006;174:437–445.

22 Xu J, Wang F, Van Keymeulen A, Herzmark P, Straight A, Kelly K, Takuwa Y, Sugimoto N, Mitchison T, Bourne HR: Divergent signals and cytoskeletal assemblies regulate self-organizing polarity in neutrophils. Cell 2003;114:201–214.

23 Rohatgi R, Ma L, Miki H, Lopez M, Kirchhausen T, Takenawa T, Kirschner MW: The interaction between N-WASP and the Arp2/3 complex links Cdc42-dependent signals to actin assembly. Cell 1999;97:221–231.

24 Van Rheenen J, Condeelis J, Glogauer M: A common cofilin activity cycle in invasive tumor cells and inflammatory cells. J Cell Sci 2009;122:305–311.

25 Malech HL, Root RK, Gallin JI: Structural analysis of human neutrophil migration. Centriole, microtubule, and microfilament orientation and function during chemotaxis. J Cell Biol 1977;75:666–693.

26 Eddy RJ, Pierini LM, Maxfield FR: Microtubule asymmetry during neutrophil polarization and migration. Mol Biol Cell 2002;13:4470–4483.
27 Niggli V: Microtubule-disruption-induced and chemotactic-peptide-induced migration of human neutrophils: implications for differential sets of signalling pathways. J Cell Sci 2003;116:813–822.
28 Xu J, Wang F, Van Keymeulen A, Rentel M, Bourne HR: Neutrophil microtubules suppress polarity and enhance directional migration. Proc Natl Acad Sci USA 2005;102:6884–6889.
29 Smith LA, Aranda-Espinoza H, Haun JB, Dembo M, Hammer DA: Neutrophil traction stresses are concentrated in the uropod during migration. Biophys J 2007;92:L58–60.
30 Keller HU, Niggli V: Colchicine-induced stimulation of PMN motility related to cytoskeletal changes in actin, α-actinin, and myosin. Cell Motil Cytoskeleton 1993;25:10–18.
31 Niggli V: Rho-kinase in human neutrophils: a role in signalling for myosin light chain phosphorylation and cell migration. FEBS Lett 1999;445:69–72.
32 Eddy RJ, Pierini LM, Matsumura F, Maxfield FR: Ca2+-dependent myosin II activation is required for uropod retraction during neutrophil migration. J Cell Sci 2000;113:1287–1298.
33 Worthylake RA, Lemoine S, Watson JM, Burridge K: RhoA is required for monocyte tail retraction during transendothelial migration. J Cell Biol 2001; 154:147–160.
34 Dinauer MC: Disorders of neutrophil function: an overview. Methods Mol Biol 2007;412:489–504.
35 Lokuta MA, Cooper KM, Aksentijevich I, Kastner DL, Huttenlocher A: Neutrophil chemotaxis in a patient with neonatal-onset multisystem inflammatory disease and Muckle-Wells syndrome. Ann Allergy Asthma Immunol 2005;95:394–399.
36 Hayward AR, Harvey BA, Leonard J, Greenwood MC, Wood CB, Soothill JF: Delayed separation of the umbilical cord, widespread infections, and defective neutrophil mobility. Lancet 1979;i:1099–1101.
37 Anderson DC, Schmalstieg FC, Shearer W, Becker-Freeman K, Kohl S, Smith CW, Tosi MF, Springer T: Leukocyte LFA-1, OKM1, p150,95 deficiency syndrome: functional and biosynthetic studies of three kindreds. Fed Proc 1985;44:2671–2677.
38 Gu Y, Williams DA: RAC2 GTPase deficiency and myeloid cell dysfunction in human and mouse. J Pediatr Hematol Oncol 2002;24:791–794.
39 Hernandez PA, Gorlin RJ, Lukens JN, Taniuchi S, Bohinjec J, Francois F, Klotman ME, Diaz GA: Mutations in the chemokine receptor gene CXCR4 are associated with WHIM syndrome, a combined immunodeficiency disease. Nat Genet 2003;34:70–74.
40 Diaz GA: CXCR4 mutations in WHIM syndrome: a misguided immune system? Immunol Rev 2005; 203:235–243.
41 Gulino AV, Moratto D, Sozzani S, Cavadini P, Otero K, Tassone L, Imberti L, Pirovano S, Notarangelo LD, Soresina R, Mazzolari E, Nelson DL, Badolato R: Altered leukocyte response to CXCL12 in patients with warts hypogammaglobulinemia, infections, myelokathexis (WHIM) syndrome. Blood 2004;104: 444–452.
42 Martin C, Burdon PC, Bridger G, Gutierrez-Ramos JC, Williams TJ, Rankin SM: Chemokines acting via CXCR2 and CXCR4 control the release of neutrophils from the bone marrow and their return following senescence. Immunity 2003;19:583–593.
43 Suratt BT, Petty JM, Young SK, Malcolm KC, Lieber JG, Nick JA, Gonzalo JA, Henson PM, Worthen GS: Role of the CXCR4/SDF-1 chemokine axis in circulating neutrophil homeostasis. Blood 2004;104:565–571.
44 Badolato R, Sozzani S, Malacarne F, Bresciani S, Fiorini M, Borsatti A, Albertini A, Mantovani A, Ugazio AG, Notarangelo LD: Monocytes from Wiskott-Aldrich patients display reduced chemotaxis and lack of cell polarization in response to monocyte chemoattractant protein-1 and formyl-methionyl-leucyl-phenylalanine. J Immunol 1998; 161:1026–1033.
45 Zicha D, Allen WE, Brickell PM, Kinnon C, Dunn GA, Jones GE, Thrasher AJ: Chemotaxis of macrophages is abolished in the Wiskott-Aldrich syndrome. Br J Haematol 1998;101:659–665.
46 Hull KM, Shoham N, Chae JJ, Aksentijevich I, Kastner DL: The expanding spectrum of systemic autoinflammatory disorders and their rheumatic manifestations. Curr Opin Rheumatol 2003;15:61–69.
47 So A, De Smedt T, Revaz S, Tschopp J: A pilot study of IL-1 inhibition by anakinra in acute gout. Arthritis Res Ther 2007;9:R28.
48 Petrilli V, Dostert C, Muruve DA, Tschopp J: The inflammasome: a danger sensing complex triggering innate immunity. Curr Opin Immunol 2007;19: 615–622.
49 Church LD, Cook GP, McDermott MF: Primer: inflammasomes and interleukin-1β in inflammatory disorders. Nat Clin Pract Rheumatol 2008;4:34–42.
50 Ting JP, Lovering RC, Alnemri ES, Bertin J, Boss JM, Davis BK, Flavell RA, Girardin SE, Godzik A, Harton JA, Hoffman HM, Hugot JP, Inohara N, Mackenzie A, Maltais LJ, Nunez G, Ogura Y, Otten LA, Philpott D, Reed JC, Reith W, Schreiber S, Steimle V, Ward PA: The NLR gene family: a standard nomenclature. Immunity 2008;28:285–287.

51 Muruve DA, Petrilli V, Zaiss AK, White LR, Clark SA, Ross PJ, Parks RJ, Tschopp J: The inflammasome recognizes cytosolic microbial and host DNA and triggers an innate immune response. Nature 2008;
52 Yu JW, Fernandes-Alnemri T, Datta P, Wu J, Juliana C, Solorzano L, McCormick M, Zhang Z, Alnemri ES: Pyrin activates the ASC pyroptosome in response to engagement by autoinflammatory PSTPIP1 mutants. Mol Cell 2007;28:214–227.
53 Papin S, Cuenin S, Agostini L, Martinon F, Werner S, Beer HD, Grutter C, Grutter M, Tschopp J: The SPRY domain of Pyrin, mutated in familial Mediterranean fever patients, interacts with inflammasome components and inhibits pro-IL-1β processing. Cell Death Differ 2007;14:1457–1466
54 Shoham NG, Centola M, Mansfield E, Hull KM, Wood G, Wise CA, Kastner DL: Pyrin binds the PSTPIP1/CD2BP1 protein, defining familial Mediterranean fever and PAPA syndrome as disorders in the same pathway. Proc Natl Acad Sci USA 2003;100:13501–13506.
55 Chae JJ, Wood G, Masters SL, Richard K, Park G, Smith BJ, Kastner DL: The B30.2 domain of pyrin, the familial Mediterranean fever protein, interacts directly with caspase-1 to modulate IL-1β production. Proc Natl Acad Sci USA 2006;103:9982–9987.
56 Wise CA, Gillum JD, Seidman CE, Lindor NM, Veile R, Bashiardes S, Lovett M: Mutations in CD2BP1 disrupt binding to PTP PEST and are responsible for PAPA syndrome, an autoinflammatory disorder. Hum Mol Genet 2002;11:961–969.
57 Huttenlocher A, Frieden IJ, Emery H: Neonatal onset multisystem inflammatory disease. J Rheumatol 1995;22:1171–1173.
58 Goldfinger SE: Colchicine for familial Mediterranean fever. N Engl J Med 1972 287:1302.
59 Dinarello CA, Wolff SM, Goldfinger SE, Dale DC, Alling DW: Colchicine therapy for familial mediterranean fever. A double-blind trial. N Engl J Med 1974;291:934–937.
60 Cooper KM, Bennin DA, Huttenlocher A: The PCH family member PSTPIP1 targets to the leukocyte uropod and regulates directed cell migration. Mol Biol Cell 2008;19:3180–3191.
61 McDermott MF, Aksentijevich I, Galon J, McDermott EM, Ogunkolade BW, Centola M, Mansfield E, Gadina M, Karenko L, Pettersson T, McCarthy J, Frucht DM, Aringer M, Torosyan Y, Teppo AM, Wilson M, Karaarslan HM, Wan Y, Todd I, Wood G, Schlimgen R, Kumarajeewa TR, Cooper SM, Vella JP, Amos CI, Mulley J, Quane KA, Molloy MG, Ranki A, Powell RJ, Hitman GA, O'Shea JJ, Kastner DL: Germline mutations in the extracellular domains of the 55 kDa TNF receptor, TNFR1, define a family of dominantly inherited autoinflammatory syndromes. Cell 1999;97:133–144.

Prof. Anna Huttenlocher
Departments of Pediatrics and Medical Microbiology and Immunology
University of Wisconsin-Madison, 4205 Microbial Sciences Building
1550 Linden Dr., Madison, WI 53706 (USA)
Tel. +1 608 265 4642, Fax +1 608 262 8418, E-Mail huttenlocher@wisc.edu

Entschladen F, Zänker KS (eds): Cell Migration: Signalling and Mechanisms.
Transl Res Biomed. Basel, Karger, 2010, vol 2, pp 40–53

Coordination of Leukocyte Polarity and Migration

Noa B. Martín-Cófreces[a,b] · Juan M. Serrador[b] · Francisco Sánchez-Madrid[a,b]

[a]Servicio de Inmunología, Hospital de La Princesa-Universidad Autónoma de Madrid, and [b]Fundación Centro Nacional de Investigaciones Cardiovasculares, Madrid, Spain

Abstract

Many important physiological processes in the body require coordinated and guided cell movement. This is exemplified during fetal development, when cells move and differentiate to build the body. In adults, cell movement is a tightly regulated event essential for tissue repair and for the homeostasis and function of the immune system. Cells detect different stimuli in their environment and are able to organize their movement based on the direction and concentration gradients of the stimulatory molecules, a process called chemotaxis. To respond to chemotactic signals, the cell must polarize, acquiring and maintaining a spatial and functional asymmetry. A large body of evidence indicates that cell polarity is essential for cell migration during leukocyte-mediated immune responses. In contrast, during chemokinesis, cells can respond to chemoattractants in a random manner, in which the cell does not follow a given path. Chemotaxis and chemokinesis both require cell polarization as the basis of their movement. Whether these two kinds of cell motility are regulated by the same molecular mechanisms is still an open question.

Leukocyte Polarization

When a leukocyte polarizes, it adopts an asymmetrical shape, with distinct morphological areas that determine its movement. Two discrete poles are then well defined: the leading edge at the front of the cell, and the uropod or trailing edge at the rear (fig. 1) [1, 2]. The maintenance of polarity is essential for all movement. Polarity is achieved by the differential recruitment of chemoattractant and specific adhesion receptors (chemokine and integrin receptors) at the front of the cell, where actin-rich structures such as filopodia, pseudopodia and the leading lamella protrude (fig. 2). In contrast, the microtubule-organizing center (MTOC), Golgi apparatus (GA) and associated vesicles are directed backwards to the uropod (fig. 2). Much work is

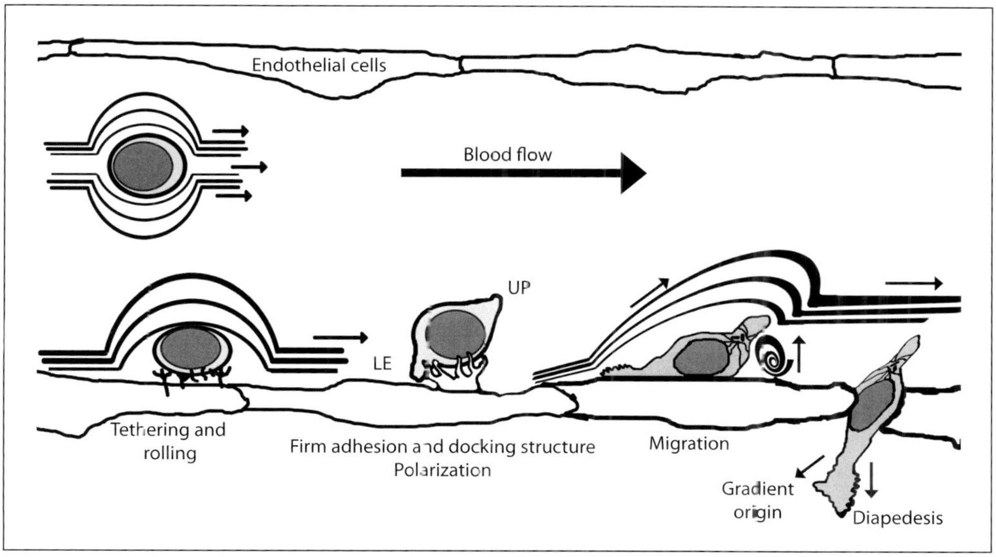

Fig. 1. A circulating leukocyte moves along a blood vessel under hemodynamic flow. The circulating cell is globular, which is the optimum geometrical form for minimizing drag. When the circulating leukocyte senses a chemoattractant gradient, it rolls over the endothelium thanks to the interaction of selectin proteins with their ligand, and comes to a stop over the endothelium through the generation of firm adhesions through integrins. To prevent dispersion of the chemoattractant, chemokines are held on the endothelial cell surface by extracellular matrix molecules and proteoglycans. In this way, the target site for leukocyte extravasation (diapedesis) is clearly marked in the wall of the blood vessel. Once attached through the docking structure, the leukocyte adopts an elongated form to allow its movement over the activated endothelium (polarization), even against the blood flow. The leading edge (LE) is a thin protrusion at the front of the cell that is closely attached to the endothelial cell surface, and shows low resistance to blood flow. The leukocyte body has a fluid-dynamic section and the uropod (UP) at the trailing edge. Its design provides maximum downforce with a minimum of drag inside the continually moving fluid. To follow the chemokine gradient to its origin, the leukocyte must leave the circulatory system and cross the endothelial layer, and the polarized shape of the leukocyte enables it to slip between the plasma membranes of adjacent cells.

needed to define the specific mechanisms for this segregation in lymphoid cells, since non-lymphoid cells, such as fibroblasts, polarize their MTOC and GA towards the leading edge, in front of the cell nucleus [1, 2]. Indeed, in polarized leukocytes most of the adhesion molecules, the TcR and the co-stimulatory receptors are segregated to the cell rear (fig. 2). This differential organization allows the formation of new attachments to the substratum at the front, where β_2-integrins are directed. The release of former attachments is detected at the rear of the cell, where the components are directed to recycling compartments near the MTOC, and redirected to the leading edge through membrane traffic via early endosomes [3–5]. In this review we focus on the requirements for correct cell polarization in leukocytes and the consequences of this process on migration.

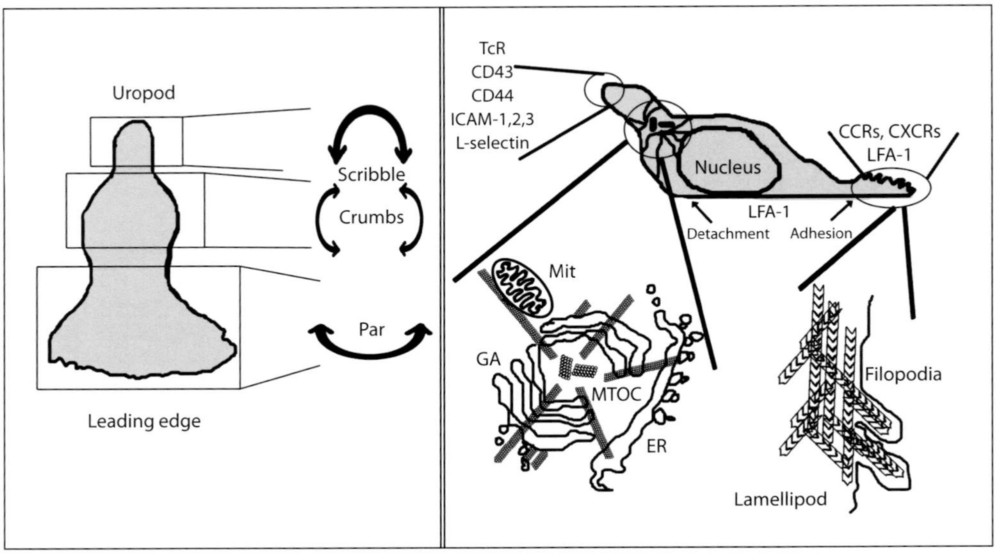

Fig. 2. Three areas or zones can be distinguished in a polarized leukocyte. (i) The leading edge at the front has a prominent lamellipod and filopodia at its end to scan the environment in the search of chemoattractants and integrin ligands. Actin nucleation and branching permits the formation of these structures. Adhesion to substrate is maximal at this zone, and actin moves in a retrograde flux over the integrin attachments. (ii) The midbody is where the tracking forces over the integrin arrays are at their strongest, to allow detachment from the substrate. (iii) The uropod at the rear contains the MTOC, mitochondria (Mit), Golgi (GA), endoplasmic reticulum (ER) and associated vesicles on the inside, and the segregated TcR, co-stimulatory and adhesion receptors on the plasma membrane. In mammals, these three areas can be distinguished by the presence of three major polarity complexes: the Par complex at the leading edge, Crumbs at the midbody, near the base of the uropod, and Scribble at the uropod.

Cell migration involves a continuous and cycling flux of membrane from the front towards the uropod and vice versa, recycling components in order to restore the molecules that allow forward movement. To achieve directional movement, the adhesion components that adhere the cell to the substratum must be interconnected to generate tracking forces. To produce a net forward movement, these forces must be minimal at the leading edge, to allow spreading over the substratum, and maximal at the rear, promoting cell detachment (fig. 2). The recycling of attachments at the rear favors this process, but a main process involved here is the control of cytoskeleton components [6, 7]. Some of the mechanisms by which the cytoskeleton controls cell polarity and migration are the active polymerization of actin at the leading edge; the contractile activity of the actomyosin cytoskeleton (mainly directed towards the uropod and midbody of polarized cells: retrograde actomyosin flow is made up of the sum of all events), and the pulling forces generated by dynein/dynactin minus-end-directed motor complexes through their interaction with radial microtubules arising from the MTOC at the uropod [4]. In addition, the cytoskeleton provides the molecular basis for intracellular membrane trafficking: vesicles and other cargoes are moved along actin and tubulin cables inside the cell. Using these tools, the cytoskeleton organizes actin-rich protuberances at the leading edge, such as lamellae and filopodia, as well as the trafficking center at the rear, which is organized around the MTOC. Moreover, the cytoskeleton and associated proteins are also needed for the maintenance of receptors in discrete domains or platforms at the plasma membrane. Overall, the cell cytoskeleton allows the polar distribution of components inside the cell, and establishes the molecular basis of the asymmetry essential for cell movement [6].

Sensing Chemotactic Gradients

Two large groups of chemoattractant molecules have been defined for eukaryotic cells: (i) ligands for tyrosine kinase receptors and (ii) chemokines, molecules that bind to seven-transmembrane receptors coupled to heterotrimeric G proteins (GPCR). Chemokines are a family with over 50 members, acting through more than 20 GPCRs, and represent the most important chemotactic molecules involved in the recruitment of immune cells [8]. Although most chemokines are secreted molecules, in vivo they are probably presented to circulating leukocytes bound to various extracellular matrix molecules and cell surface proteoglycans (fig. 1) [9]. The major roles of chemokines are to activate integrins through inside-out signaling, to reinforce leukocyte adherence and to induce chemotaxis in various tissue microenvironments [9]. Chemokine receptors are segregated to the leading edge of migrating lymphocytes. Specificity in leukocyte migration is regulated at multiple levels, since different receptors may bind to and be activated by several chemokines. There is differential tissue expression of chemokines and adhesion molecules, and a corresponding limited and specific expression of chemokine receptors in different leukocyte subsets. Finally, combinatorial expression of chemokine receptors and adhesion molecules makes leukocyte migration more specific [9]. In the homing of T and B cells to lymph node and spleen, it has been described that the differential localization of these lymphocytes to specific areas is dependent on the subset of chemokine receptors exposed in their membranes. T cells are attracted via CCR7 towards their specific destination by a CCL19 and CCL21 gradient, whereas B cells follow a CXCL13 gradient sensed by CXCR5 until they reach the B-cell follicle [10]. The binding of a chemokine by its receptor promotes the dissociation of Gαβγ trimer in free activated Gα and the Gβγ dimer. Both control the activity of ion channels and enzymes, promoting an intracellular increase in Ca^{2+} flux [11]. Downstream of chemokine receptors is the Rho small GTPase family, formally divided into seven subfamilies: Rho, Rac, Cdc42, RhoD, RhoG, RhoE and TC10. These proteins have specific roles in actomyosin cytoskeleton remodeling, and belong to the signaling pathways most analyzed in polarization events [12].

The activity of the GTPase RhoA is necessary to increase the adhesiveness of integrins upon chemokine binding to its receptor. RhoA controls the contractile activity of the actomyosin network through myosin light chain (MLC) phosphorylation by ROCK protein kinase [13]. Indeed, MLC kinase (MLCK) increases the cellular content of phosphorylated MLC, and its activity is dependent on calcium-calmodulin (Ca^{2+}/CaM) binding [13]. Rac1 activity is also important for controlling the extent of MLC phosphorylation through the activation of the p21-activated protein kinase (PAK), which in turn phosphorylates and therefore inhibits MLCK [13]. RhoA may also promote the MLC phosphorylation indirectly, through specific interaction with the myosin-binding subunit of myosin phosphatase (MLCP). ROCK-dependent phosphorylation of MLCP allows an increase in the extent of phosphorylation of MLC. Once phosphorylated at Ser19, MLC promotes myosin heavy chain ATPase activity, providing the basis for F-actin contraction [14]. Actin-myosin assembly increases the stability of the actin cytoskeleton and the sliding of F-actin, favoring the existence of tensional forces on integrin-mediated cell attachments to the substratum. Through this process, the actomyosin network acts as a tension cable over the rearward integrin array in contact with the substratum, a function essential for efficient cell migration [2]. Both RhoA and Myosin II are located at the rear and side of polarized leukocytes, where they promote cell body contraction and inhibit Rac activity to avoid membrane protrusions at this location [12]. In contrast, Rac1 activity at the leading edge would inhibit myosin contraction through MLCK inhibition. It has been shown that myosin II activated by MLCK at the cell periphery controls membrane ruffling, whereas ROCK would phosphorylate MLC preferentially at the center of the migrating cell. MLCP action at the cell periphery allows correct membrane ruffling. Therefore, the spatial regulation of MLC phosphorylation plays critical roles in controlling cell migration [15]. In relation to these findings, myosin IIA has been shown to regulate the interaction of LFA-1 ($\alpha_L\beta_2$) integrin with its ligand (ICAM-1). The activity of Myosin IIA is then necessary to promote LFA-1 detachment and, concomitantly, the retraction of the uropod [16]. In this regard, the use of the MLCK (myosin light chain kinase) inhibitors ML-7 and ML-9 during the stimulation of neutrophils with the chemoattractant *N*-formyl-l-methionyl-l-leucyl-l-phenylalanine prevents cell polarization but allows the cell to spread, resulting in a broad lamellipod around the cell perimeter [17]. Myosin IIA has also been reported to localize at the leading edge of lymphocytes, where it interacts with the chemokine receptor CXCR4 and participates in the endocytosis of the receptor upon ligand binding [18]. Therefore, myosin IIA participates in cell polarity at various levels to allow the movement of a given cell along a chemokine gradient. In physiological scenarios where the chemokine is not abundant in the vicinity of a cell, chemokine receptors are engaged at the leading edge, which is formed oriented toward the intercellular space where the chemokine is enriched. The uropod retracts at the rear of the cell, and the leukocyte shows a net movement directed to the source of the chemokine gradient. Myosin will help to clear the plasma membrane of engaged CXCR4 molecules, favoring their recycling, and will promote the retrograde actin flow that allows integrin detachment [18, 19]. Interestingly, phosphorylated MLC was found both at the leading and trailing edges during lymphocyte migration, and mitochondria were localized around the uropod-polarized MTOC in these cells [20]. The MLC phosphorylation detected at the uropod was shown to be dependent on the polarization and activity of mitochondria, whereas phosphorylation at the cell front was not affected by F0/F1 ATPase inhibitors. During lymphocyte polarization, mitochondrial dynamics are essential. Mitochondrial fission, driven by dynamin-related protein 1 (Drp1), promotes organelle relocation and supports lymphocyte chemotaxis, whereas mitochondrial fusion, controlled by Opa1 and various other proteins, inhibits both processes [20].

Establishment of Two Poles: Front-Back Coordination

Rho GTPase activity is also necessary to promote the phosphorylation of ERM (ezrin-radixin-moesin) proteins at the trailing edge [21]. ERM proteins act as linkers between the plasma membrane and the actin cytoskeleton, thereby regulating cell deformability. ERM proteins are thus responsible for cell adhesion and cortical cytoskeleton morphogenesis [22]. ERM proteins interact with adhesion molecules at the uropod of migrating lymphocytes, such as ICAMs-1, -2 and -3, L-selectin, CD43 and CD44, and connect these proteins to the actin cortical cytoskeleton. The polarization of adhesion receptors to the uropod is necessary for correct migration, and the uropod has been proposed as a reservoir for these molecules, preventing their adhesive function during migration to avoid the cell from getting stuck to the substratum [19]. Dominant negative forms of RhoA, Rac1 and Cdc42 GTPases promote uropod formation, with concomitant accumulation of moesin in non-polarized cells, whereas activated mutants of these proteins prevented moesin polarization in constitutively polarized cells [23]. During cell migration, Rac and Cdc42 respectively orchestrate the organization of lamellae and filopodia, together with other protrusive structures, at the leading edge. The overexpression of mutant forms of either Rac or Cdc42 in T lymphocytes results in loss of polarity and impaired migration towards gradients of the chemokine SDF-1α [23]. Rac and Cdc42 both regulate the activity of the Arp2/3 complex at the cell front, favoring actin nucleation at given cellular locations and the branching of actin filaments through WASP (Wiskott-Aldrich syndrome protein) and WAVE (WASP-family verprolin-homologous protein) [7].

However, the situation appears to be more complex than this. Rac1 activity has also been detected at the trailing edge of live migrating neutrophils, where it is necessary for the correct retraction of the uropod [24]. Neutrophils from Rac2-null mice have impaired chemotaxis due to a marked defect in lamellipodia formation, whereas *rac1*$^{-/-}$ cells form multiple unstable lamellipodia and develop an elongated morphology due to a uropod retraction defect [25]. Moreover, new players have recently been identified in the acquisition and maintenance of leukocyte polarity. One of these is the Ras-like GTPase Rap1, whose function in inside-out signaling and integrin-mediated matrix adhesion is important for the establishment of T-lymphocyte polarity and dendritic and T-cell migration. In these events, the Rap1 effector protein RAPL is of special importance in the modulation of β_1, β_2 and β_3 integrins [26]. RapL binds active Rap1 and mediates the clustering of LFA-1 integrin and its adhesion to its ligand ICAM-1 [26]. Indeed, upon overexpression of a GTPase-activating protein specific for Rap1, the decrease in Rap-1-dependent signaling impairs chemokine-induced polarization and transendothelial migration. This last effect is due to a defect in the binding of LFA-1 to ICAM-1 [26]. The expression of a truncated RAPL mutant unable to bind to Rap1 (RAPL ΔN) abrogates the polarizing effect of the constitutively active mutant V12Rap1 and blocks chemokine-induced T-cell polarization [26]. Nevertheless, little is known about the signaling pathways used by Rap1 and chemokines to induce T-cell polarization. A recent study has shown that a number of polarity proteins (e.g., Par3, aPKC, Scribble, Dlg, and Crumbs3) are differentially segregated in polarized T cells [27], suggesting that they might regulate T-cell polarization. To date, three polarity complexes have been described: partitioning defective (Par), Scribble, and Crumbs [28]. The polarization of Scribble and Crumbs complexes in the uropod is necessary to maintain the restricted localization of ezrin and CD44 in polarized T cells. Crumbs3 is concentrated at the base of the uropod, near the midbody, whereas Scribble complex is segregated to the area of CD44 clustering. Consequently, prevention of Scribble expression provokes defects in cell migration [29]. Par complex seems to be excluded from the areas where Scribble and Crumbs are influential. It has been recently shown that Rap1 activity at the leading edge during SDF-1α-mediated chemotaxis of T lymphocytes is regulated by TIAM-1 (T lymphoma invasion and metastasis 1), a guanosine nucleotide-exchange factor (GEF) for Rac. Tiam1, in conjunction with the Par polarity complex (consisting of a core of Par3 and Par6 proteins and atypical PKCs (aPKCλ/ι and aPKCζ)), might transform the activity of Rap1 protein into Rac1 activity. This would rely on Rap-1-dependent activation of Cdc42, which would then activate the atypical PKCs. The PKC-mediated phosphorylation of Tiam1 triggers its GEF activity, thus increasing GTP-Rac and promoting the activity of Rac effectors [28]. In addition, Par3 and Par6 proteins have been implicated in the targeting of RhoA to the proteasome for ubiquitin-dependent degradation [29]. RhoA degradation at the leading edge would reduce actin contractility at lamellipodia and filopodia, allowing sprouting of the membrane. Therefore, the balance between the Rho activity and the activities of Rac and Cdc42 constitutes a mechanism for fine tuning leukocyte polarity, and is tightly regulated through a variety of molecular pathways.

The described role of Tiam1 activity does not completely explain the degree of Rac1 activation observed upon the binding of SDF-1α to its receptor. In fact, two other GEFs for Rac1, DOCK2 and Vav1, have been analyzed extensively in this context. DOCK2 and Vav1 are both activated by the binding of SDF1-α to CXCR4 and localize to the leading edge of polarized lymphocytes [11, 30]. DOCK2 has also been analyzed in migration events stimulated by other chemokines, such as CCL21 and CCL19, and its role in T- and B-lymphocyte migration has been clearly established [11]. In both lymphocyte types, Rac1 activity was dependent on the presence of these GEFs, and defects in Rac activation correlated with defects in actin polymerization and the increase in integrin adhesiveness to its ligands, finally preventing lymphocyte migration [11, 30]. Furthermore, Rap1 activity in response to chemokine stimulation is defective in Dock2-deficient mice [11], a finding which correlates well with the Tiam1-dependent model for Rac1 activation described above [28]. Aside from T and B lymphocytes, Dock2 seems also to be necessary for the chemotaxis of plasmacytoid dendritic cell [31]; but in contrast it appears to be dispensable for the migration of monocytes [11, 30] and myeloid dendritic cells [31]. Much work remains to be done to decipher the specific requirements of each leukocyte type in polarization and migration.

The small GTPase Cdc42 has been shown to be redistributed to the leading edge of motile leukocytes, where it probably occurs in its activated form [2]. Cdc42 activity was shown to be dependent on the activity of αPix. Members of the Pix protein family have been shown to be important GEFs for Rac and Cdc42. The predominant binding partners for PIX proteins, however, are GIT proteins 1 and 2 (G-protein-coupled receptor kinase-interacting proteins), also known as CAT proteins, p95PKL or APP1/2 [32]. PIX and GIT proteins associate in large, stable oligomeric complexes that recruit Rac1 and Cdc42 GTPases and PAK kinases [32]. The mammalian PDZ-containing Scribble complex interacts with the C-terminal domain of βPix, thus bringing it into contact with GIT1, and this complex has been linked to the recycling of GPCRs in non-lymphoid cells [32]. However, the Scribble complex is directed to the uropod in motile T lymphocytes, as discussed above. Thus, Cdc42 and Rac activity might be controlled at the trailing edge by the Scribble polarity complex and at the leading edge by the Par complex. These multiproteic interactions would then be in charge of the compartmentalization of Cdc42 and Rac activities at the two poles of a polarized lymphocyte through the relation with different partners and effectors.

Microtubule Connection: The Lost Link?

It is possible that both Cdc42 and Rac connect to microtubules (MTs) through their binding to IQGAP1 (IQ motif containing GTPase-activating protein 1) and CLIP170 (cytoplasmic linker protein-170). Through these proteins, Cdc42 and Rac would capture microtubule plus-end tips at the plasma membrane [33], and this could serve to link the radial array of MTs to the cortical cytoskeleton at the uropod. In polarized leukocytes the MTOC localizes at the uropod, behind the cell nucleus [3]. A similar polarized array of MTs has been observed in neutrophils during chemotaxis

[34]. Because the relative rigidity of MTs would limit cell deformability, the polarization of the MT array to the uropod would streamline cell shape and facilitate passage through narrow collagen matrices and endothelial monolayers [35]. Interestingly, the MTOC of polarized lymphocytes is enriched in acetylated MTs, and this concentration of these stable MTs at the uropod might be a mechanism to allow increased cell softness during migration; the MTOC-related array of acetylated MTs might thus be important for cell shape and maintenance of the uropod [36]. The scaffolding function of histone-deacetylase 6 (HDAC6), which regulates the acetylation of MTs at Lys40 of α-tubulin, was found to be important for lymphocyte motility, even though its deacetylase activity towards MTs is not important during T-cell polarization [36]. The use of specific MT polymerization inhibitors to study the role of MTs during cell migration has yielded contradictory results. Colchicine and nocodazole have been variously reported to stimulate random migration of human neutrophils, to have no effect or to mildly inhibit random migration when used at high concentrations [37]. Related studies using these drugs to address the role of MTs in chemotaxis yielded similarly conflicting results, with reorientation towards the chemoattractant focus being either impaired or not significantly affected [37]. The issue is further complicated by the fact that MT-inhibiting drugs such as nocodazole trigger an F-actin-dependent cell polarization in polymorphonuclear cells in the absence of chemoattractant, probably due to Rho GTPase activation [37]. In light of various observations, it is conceivable that the reorientation of the MT array to the uropod might act to strengthen polarity once it has been established. The question then remains as to how the asymmetric MTs might affect cell polarity during migration. One possibility is that the MT array might contribute to the maintenance of cell polarity by modulating the activity of Rho family GTPases, key regulators of actin dynamics and organization [38]. Consistent with this possibility, the colchicine-related spontaneous polarization of neutrophils is prevented by inhibition of Rho-GTPase, suggesting that disruption of MTs might concomitantly stimulate Rho activity and myosin II activity through MLC phosphorylation [39]. In this regard, Rho/ROCK/Myosin II-dependent polarization of clathrin-based structures to the uropod is necessary for a correct migration. Analysis of clathrin function at the uropod shows that although these vesicles and the MTs localize to analogous areas in polarized lymphocytes, colchicine treatment still allows clathrin-mediated traffic, suggesting that the MT array, which is important for long-range movement of membrane vesicles, is dispensable. Colchicine, through the activation of Rho GTPase, might potentiate the clathrin-mediated traffic. Further research is necessary to assess MT role in this function [40].

The study of MT plus-end turnover in non-lymphoid cells has shown that the LPA/Rho-mDia signaling module stabilizes MTs through the capping of plus ends, thereby preventing tubulin subunit exchange [33]. The molecular mechanisms that sustain this process are still poorly understood. Rho/mDia-stabilized MTs show increased levels of detyrosinated α-tubulin, which is posttranslationaly modified through the removal of the C-terminal tyrosine residue, exposing a glutamate residue at the C terminus; in contrast, more dynamic MTs contain tyrosinated α-tubulin [33]. The current model for mDia-regulated MT stabilization suggests that LPA triggers a Rho/mDia-dependent pathway that also involves GSK3β, the plus-end tracking proteins EB1 and APC, and novel PKCs to generate a polarized array of stabilized MTs [33, 41]. The localization of these stabilized MTs is regulated by integrin signaling [33], which is thought to contribute to cell polarity by directing vesicular trafficking or actin regulators to the leading edge. More recently, actin nucleation and MT stabilization by mDia2 have been found to be independent of mDia2 dimerization. Moreover, purified mDia2, through its FH2 domain, directly stabilizes MTs by reducing the rates of both polymerization and depolymerization [42]. Finally, the actin and tubulin cytoskeletons are involved in regulating leukocyte rigidity and deformability, but the major regulator of this function has been considered to be the vimentin-based intermediate filaments. However, lymphoid cells from *vim*$^{-/-}$ mice polarize correctly, suggesting that vimentin is dispensable for T-cell polarization [43]. AKAP450 (a kinase-anchoring protein of 450 kDa) is a scaffolding protein important for LFA-1-induced T lymphocyte motility and polarity. The triggering of this specific integrin provokes the recruitment of AKAP450 along the MTs arising from the MTOC. The presence and function of AKAP450 is important for the recruitment of a signaling complex formed by LFA-1 and PKCs β and δ, and which also includes tubulin. AKAP450 is tightly associated to the MTOC and the Golgi complex [44]. Recently, a tubulin-nucleating activity potentially important for lymphocyte polarization has been described for AKAP450 and associated GM130 at the *cis*-side of the Golgi complex in non-lymphoid cells [45]. Migration of epithelial cells in wound-healing experiments was defective in AKAP450-depleted cells, but no defects in MTOC polarization were found. Instead, the authors found that short MTs arising from AKAP450 and γTuRC complexes (γ-tubulin ring complex, important for tubulin nucleation) were covered by CLASP2 protein, and that dynein/dynactin complexes are important for the anchorage of Golgi-arising MTs. Little is known about microtubule nucleation in lymphocyte migration and polarization, but it is clear that structural proteins and enzymes important for regulation of MT dynamics, such as AKAP450 and HDAC6, form a part of this puzzle and are important for T-cell polarity and migration. The specific role of MTs in lymphocyte polarization and migration deserves further research.

Segregated Signalling Domains in Polarized Lymphocytes

Cell membranes have been analyzed in the context of lymphocyte polarization. In contrast with fibroblast plasma membrane at the leading edge, which is enriched in both GM1 and GM3 ganglioside raft subtypes, T lymphocytes and neutrophils have been found to show a polarization of GM3 rafts to the leading edge (L-raft) and of GM1 rafts to the uropod (U-raft) [12]. Emerging questions are how the cell might sense and segregate these kinds of lipids and whether this segregation is a strategy used by polarized lymphoid cells to differentiate molecules destined for recycling to the uropod or the leading edge. Moreover, forces generated by myosin II also contribute to the redistribution of large-scale, detergent-resistant membrane domains to the uropod after PMN polarization, perhaps serving to partition key molecules essential for the specific functions of the lamellipod and uropod [46]. The inositide composition of membranes might be relevant to intracellular signaling. Upon chemoattractant binding, the PI3K family of enzymes are activated and catalyze the conversion of phosphatidyl inositol 4,5-biphoshate (PIP_2) to phosphatidyl inositol 3,4,5-triphosphate (PIP_3), through phosphorylation at the 3′ position of the inositol ring. PI3Ks are usually heterotrimeric proteins, formed by the combination of catalytic and regulatory subunits to form three major classes of enzymes [47]. In leukocytes it is not clear whether PIP_3 production is restricted to the leading edge as in *Dictyostelium*, where PI3K and PTEN, the corresponding phosphatase, are localized at the leading and trailing edge, respectively [7]. PI3Kγ has been shown to affect neutrophil and macrophage migration, but shows only subtle effects on T- and B-cell polarization and migration. In contrast, PTEN activity increases T- and B-lymphocyte motility [12]. It is possible that other phosphatases act at this step, such as SHIP1, which was found to be essential for neutrophil polarization and chemotaxis, in contrast with the weak action of PTEN in similar experiments [48]. However, more recent studies suggest that PTEN may be important for neutrophil directed movement by acting as a sensor that prioritizes among a hierarchy of chemotactic gradients in the medium to produce a focused response [49].

In addition, the cell content and localization of PIP_2, a direct regulator of many actin-binding and remodeling proteins, including GTPases [50], might be of crucial importance for polarized signaling and molecular localization. At the leading edge, PIP_2 could be the substrate for either PI3K (as described above) or PLC (phospholipase C). PLC hydrolyzes the molecule to produce inositol 1,4,5-triphosphate (I3P), which is essential for increases in Ca^{2+} intracellular flux, and diacylglycerol, which activates PKC [51]. As described above, polarity complexes, such as Par at the leading edge, contain several PKCs that are important for T-cell polarity and chemotaxis [27]. This agrees with findings showing that PLC is essential for Ca^{2+}-independent, DAG-dependent T-cell chemotaxis towards CCL17 and CCL12, ligands of the chemokine receptor CCR4 [52]. PIP_2 is also important at the uropod for the activation of ERM proteins during leukocyte chemotaxis [4]. The binding of ERMs to PIP_2 and their subsequent phosphorylation at C-terminal Ser/Thr residues enables cross-linking of several receptors to the actin cytoskeleton at the uropod, promoting their polarization. The phosphorylation of ERMs is essential for T-cell polarization and migration [4]. The composition and regulation of membrane dynamics are therefore important for leukocyte polarization, and this is still a field of intense research. Recent studies on type 1 phosphatidylinositol-4-phosphate-5-kinase (PIP5K1) has shown that this enzyme, responsible for PIP_2 synthesis, is localized at the uropod of migrating neutrophils, where it contributes to uropod retraction, and is in charge of maintaining a polarized shape and chemotaxis [4].

Concluding Remarks

Leukocyte polarization is a tightly regulated process, with important physiological roles in immune responses. Immune cells need to polarize to allow correct locomotion and to ensure migration along guided pathways that are generated by specific chemoattractants. This polarization requires expression and regulation of the appropriate surface receptors and intracellular signaling and cytoskeletal components. Although important advances have been made in our understanding of these processes, some important questions remain. The importance of the microtubule cytoskeleton and its relation to the specific intracellular trafficking of membrane and organelles is still not fully defined. In addition, a more precise definition of the relative contribution of signaling pathways to polarization will require analysis of the specific composition of structural and signaling domains at the plasma membrane and the specific role of PDZ-containing proteins. The analysis of polarity complexes is an emerging field in the study of polarization mechanisms, and promises to lead to greater understanding of the differential roles of GTPases located at the leading edge and the uropod.

References

1 Sanchez-Madrid F, del Pozo MA: Leukocyte polarization in cell migration and immune interactions. EMBO J 1999;18:501–511.

2 Ridley AJ, Schwartz MA, Burridge K, Firtel RA, Ginsberg MH, Borisy G, Parsons JT, Horwitz AR: Cell migration: integrating signals from front to back. Science 2003;302:1704–1709.

3 Serrador JM, Nieto M, Sanchez-Madrid F: Cytoskeletal rearrangement during migration and activation of T lymphocytes. Trends Cell Biol 1999; 9:228–233.

4 Sanchez-Madrid F, Serrador JM: Bringing up the rear: defining the roles of the uropod. Nat Rev Mol Cell Biol 2009;10:353–359.

5 Del Pozo MA, Sanchez-Mateos P, Sanchez-Madrid F: Cellular polarization induced by chemokines: a mechanism for leukocyte recruitment? Immunol Today 1996;17:127–131.

6 Pollard TD, Borisy GG: Cellular motility driven by assembly and disassembly of actin filaments. Cell 2003;112:453–465.

7 Vicente-Manzanares M, Sanchez-Madrid F: Role of the cytoskeleton during leukocyte responses. Nat Rev Immunol 2004;4:110–122.

8 Mellado M, Rodriguez-Frade JM, Mañes S, Martinez AC: Chemokine signaling and functional responses: the role of receptor dimerization and TK pathway activation. Annu Rev Immunol 2001;19:397–421.

9 Bromley SK, Mempel TR, Luster AD: Orchestrating the orchestrators: chemokines in control of T-cell traffic. Nat Immunol 2008;9:970–980.

10 Stein JV, Nombela-Arrieta C: Chemokine control of lymphocyte trafficking: a general overview. Immunology 2005;116:1–12.

11 Thelen M, Stein JV: How chemokines invite leukocytes to dance. Nat Immunol 2008;9:953–959.

12 Gómez-Moutón C, Mañes S: Establishment and maintenance of cell polarity during leukocyte chemotaxis. Cell Adh Migr 2007;1:69–76.

13 Somlyo AV, Phelps C, Dipierro C, Eto M, Read P, Barrett M, Gibson JJ, Burnitz MC, Myers C, Somlyo AP: Rho kinase and matrix metalloproteinase inhibitors cooperate to inhibit angiogenesis and growth of human prostate cancer xenotransplants. FASEB J 2003;17:223–234.

14 Craig R, Woodhead JL: Structure and function of myosin filaments. Curr Opin Struct Biol 2006;16: 204–212.

15 Totsukawa G, Wu Y, Sasaki Y, Hartshorne DJ, Yamakita Y, Yamashiro S, Matsumura F: Distinct roles of MLCK and ROCK in the regulation of membrane protrusions and focal adhesion dynamics during cell migration of fibroblasts. J Cell Biol 2004;164:427–439.

16 Morin NA, Oakes PW, Hyun YM, Lee D, Chin YE, King MR, Springer TA, Shimaoka M, Tang JX, Reichner JS, Kim M: Non-muscle myosin heavy chain IIA mediates integrin LFA-1 de-adhesion during T lymphocyte migration. J Exp Med 2008;205: 195–205.
17 Eddy RJ, Pierini LM, Matsumura F, Maxfield FR: Ca^{2+}-dependent myosin II activation is required for uropod retraction during neutrophil migration. J Cell Sci 2000;113:1287–1298.
18 Rey M, Valenzuela-Fernandez A, Urzainqui A, Yanez-Mo M, Perez-Martinez M, Penela P, Mayor F Jr, Sanchez-Madrid F: Myosin IIA is involved in the endocytosis of CXCR4 induced by SDF-1α. J Cell Sci 2007;120:1126–1133.
19 Barreiro O, de la Fuente H, Mittelbrunn M, Sanchez-Madrid F: Functional insights on the polarized redistribution of leukocyte integrins and their ligands during leukocyte migration and immune interactions. Immunol Rev 2007;218:147–164.
20 Campello S, Lacalle RA, Bettella M, Mañes S, Scorrano L, Viola A: Orchestration of lymphocyte chemotaxis by mitochondrial dynamics. J Exp Med 2006;203:2879–2886.
21 Lee JH, Katakai T, Hara T, Gonda H, Sugai M, Shimizu A: Roles of p-ERM and Rho-ROCK signaling in lymphocyte polarity and uropod formation. J Cell Biol 2004;167:327–337.
22 Mangeat P, Roy C, Martin M: ERM proteins in cell adhesion and membrane dynamics. Trends Cell Biol 1999;9:187–192.
23 Del Pozo MA, Vicente-Manzanares M, Tejedor R, Serrador JM, Sanchez-Madrid F: Rho GTPases control migration and polarization of adhesion molecules and cytoskeletal ERM components in T lymphocytes. Eur J Immunol 1999;29:3609–3620.
24 Sun CX, Downey GP, Zhu F, Koh AL, Thang H, Glogauer M: Rac1 is the small GTPase responsible for regulating the neutrophil chemotaxis compass. Blood 2004;104:3758–3765.
25 Pestonjamasp KN, Forster C, Sun C, Gardiner EM, Bohl B, Weiner O, Bokoch GM, Glogauer M: Rac1 links leading edge and uropod events through Rho and myosin activation during chemotaxis. Blood 2006;108:2814–2820.
26 Price LS, Bos JL: RAPL: taking the Rap in immunity. Nat Immunol 2004;5:1007–1008.
27 Ludford-Menting MJ, Oliaro J, Sacirbegovic F, Cheah ET, Pedersen N, Thomas SJ, Pasam A, Iazzolino R, Dow LE, Waterhouse NJ, Murphy A, Ellis S, Smyth MJ, Kershaw MH, Darcy PK, Humbert PO, Russell SM: A network of PDZ-containing proteins regulates T-cell polarity and morphology during migration and immunological synapse formation. Immunity 2005;22:737–748.
28 Iden S, Collard JG: Cross-talk between small GTPases and polarity proteins in cell polarization. Nat Rev Mol Cell Biol 2008;9:846–859.
29 Wang HR, Zhang Y, Ozdamar B, Ogunjimi AA, Alexandrova E, Thomsen GH, Wrana JL: Regulation of cell polarity and protrusion formation by targeting RhoA for degradation. Science 2003;302:1775–1779.
30 Vicente-Manzanares M, Cruz-Adalia A, Martín-Cófreces NB, Cabrero JR, Dosil M, Alvarado-Sánchez B, Bustelo XR, Sánchez-Madrid F: Control of lymphocyte shape and the chemotactic response by the GTP exchange factor Vav. Blood 2005;105: 3026–3034.
31 Gotoh K, Tanaka Y, Nishikimi A, Inayoshi A, Enjoji M, Takayanagi R, Sasazuki T, Fukui Y: Differential requirement for DOCK2 in migration of plasmacytoid dendritic cells versus myeloid dendritic cells. Blood 2008;111:2973–2976.
32 Hoefen RJ, Berk BC: The multifunctional GIT family of proteins. J Cell Sci 2006;119:1469–1475.
33 Watanabe T, Noritake J, Kaibuchi K: Regulation of microtubules in cell migration. Trends Cell Biol 2005;15:76–83.
34 Eddy RJ, Pierini LM, Maxfield FR: Microtubule asymmetry during neutrophil polarization and migration. Mol Biol Cell 2002;13:4470–4483.
35 Ratner S, Sherrod WS, Lichlyter D: Microtubule retraction into the uropod and its role in T-cell polarization and motility. J Immunol 1997;159:1063–1067.
36 Cabrero JR, Serrador JM, Barreiro O, Mittelbrunn M, Naranjo-Suárez S, Martín-Cófreces N, Vicente-Manzanares M, Mazitschek R, Bradner JE, Avila J, Valenzuela-Fernández A, Sánchez-Madrid F: Lymphocyte chemotaxis is regulated by histone deacetylase 6, independently of its deacetylase activity. Mol Biol Cell 2006;17:3435–3445.
37 Keller H, Niggli V, Zimmermann A: Diversity in motile responses of human neutrophil granulocytes: functional meaning and cytoskeletal basis. Adv Exp Med Biol 1991;297:23–37.
38 Wittmann T, Waterman-Storer CM: Cell motility: can Rho GTPases and microtubules point the way? J Cell Sci 2001;114:3795–3803.
39 Niggli V: Microtubule-disruption-induced and chemotactic-peptide-induced migration of human neutrophils: implications for differential sets of signalling pathways. J Cell Sci 2003;116:813–822.
40 Samaniego R, Sanchez-Martin L, Estecha A, Sanchez-Mateos P: Rho/ROCK and myosin II control the polarized distribution of endocytic clathrin structures at the uropod of moving T lymphocytes. J Cell Sci 2007;120:3534–3543.

41 Eng CH, Huckaba TM, Gundersen GG: The formin mDia regulates GSK3β through novel PKCs to promote microtubule stabilization but not MTOC reorientation in migrating fibroblasts. Mol Biol Cell 2006;17:5004–5016.

42 Bartolini F, Moseley JB, Schmoranzer J, Cassimeris L, Goode BL, Gundersen GG: The formin mDia2 stabilizes microtubules independently of its actin nucleation activity. J Cell Biol 2008;181:523–536.

43 Brown MJ, Hallam JA, Colucci-Guyon E, Shaw S: Rigidity of circulating lymphocytes is primarily conferred by vimentin intermediate filaments. J Immunol 2001;166:6640–6646.

44 El Din El Homasany BS, Volkov Y, Takahashi M, Ono Y, Keryer G, Delouvee A, Looby E, Long A, Kelleher D: The scaffolding protein CG-NAP/AKAP450 is a critical integrating component of the LFA-1-induced signaling complex in migratory T cells. J Immunol 2005;175:7811–7818.

45 Rivero S, Cardenas J, Bornens M, Rios RM: Microtubule nucleation at the *cis*-side of the Golgi apparatus requires AKAP450 and GM130. EMBO J 2009;28:1016–1028.

46 Seveau S, Eddy RJ, Maxfield FR, Pierini LM: Cytoskeleton-dependent membrane domain segregation during neutrophil polarization. Mol Biol Cell 2001;12:3550–3562.

47 Sotsios Y, Ward SG: Phosphoinositide 3-kinase: a key biochemical signal for cell migration in response to chemokines. Immunol Rev 2000;177:217–235.

48 Nishio M, Watanabe K, Sasaki J, Taya C, Takasuga S, Iizuka R, Balla T, Yamazaki M, Watanabe H, Itoh R, Kuroda S, Horie Y, Forster I, Mak TW, Yonekawa H, Penninger JM, Kanaho Y, Suzuki A, Sasaki T: Control of cell polarity and motility by the PtdIns(3,4,5)P_3 phosphatase SHIP1. Nat Cell Biol 2007;9:36–44.

49 Heit B, Robbins SM, Downey CM, Guan Z, Colarusso P, Miller BJ, Jirik FR, Kubes P: PTEN functions to 'prioritize' chemotactic cues and prevent 'distraction' in migrating neutrophils. Nat Immunol 2008;9:743–752.

50 Caroni P: New EMBO members' review: actin cytoskeleton regulation through modulation of $PI_{4,5}P_2$ rafts. EMBO J 2001;20:4332–4336.

51 Rhee SG: Regulation of phosphoinositide-specific phospholipase C. Annu Rev Biochem 2001;70:281–312.

52 Cronshaw DG, Kourounalis A, Parry R, Webb A, Brown Z, Ward SG: Evidence that phospholipase-C-dependent, calcium-independent mechanisms are required for directional migration of T lymphocytes in response to the CCR4 ligands CCL17 and CCL22. J Leukoc Biol 2006;79:1369–1380.

Dr. Francisco Sánchez-Madrid
Servicio de Inmunología, Hospital de La Princesa, Planta 1
Diego de León 62, ES–28006 Madrid (Spain)
Tel. +34 91520 2370, Fax +34 9152 2374, E-Mail fsanchez.hlpr@salud.madrid.org

Entschladen F, Zänker KS (eds): Cell Migration: Signalling and Mechanisms.
Transl Res Biomed. Basel, Karger, 2010, vol 2, pp 54–66

Positioning Phosphoinositide 3-Kinase in Chemokine and Antigen-Dependent T-Lymphocyte Navigation Mechanisms

Stephen G. Ward

Department of Pharmacy and Pharmacology, University of Bath, Bath, UK

Abstract

Activation of phosphoinositide 3-kinase (PI3K) is a signaling event elicited by most chemokine receptors, contributing to actin reorganization and other components of the general migratory machinery that are necessary for T-lymphocyte homing. More recently, PI3K has also been implicated as a key regulator in novel mechanisms mediated by the T-cell antigen receptor (TCR) and the costimulatory molecule CD28, that guide the access and retention of specific T cells into antigen-rich non-lymphoid tissue. Inhibition of PI3K has therefore been proposed as a potential therapeutic strategy for T-lymphocyte-dependent pathologies such as transplant rejection as well as many autoimmune and inflammatory diseases. Here, we examine the PI3K-dependent signal transduction pathways involved in T-cell migration during distinct modes of T-cell trafficking in response to either chemokines or the TCR and/or CD28.

The migration of lymphocytes is crucial for almost all levels of T-cell biology, including their development in the thymus, entry of naive T cells into secondary lymphoid organs (SLO) and subsequent immune response initiation, maturation into circulating memory and effector T cells, followed by egress from the SLO and homing to peripheral tissues. Lymphocyte migration is coordinated by selectins, integrins and chemotactic receptors, an array of signaling events and cytoskeleton reorganization [1, 2].

Phosphoinositide 3-kinase (PI3K) has been positioned at the heart of an evolutionarily conserved cellular compass and/or the biochemical mechanisms that facilitate cell migration. PI3K has therefore become a popular drug target for inhibition of leukocyte migration in response to inflammatory chemoattractant mediators including members of the chemokine family. However, the precise role of PI3K in the regulation of cell migration remains open to refinement as numerous examples of PI3K-independent leukocyte migration (particularly with respect to T-lymphocytes) have been described. More recently, PI3K has also been implicated as a key regulator in novel mechanisms mediated by the T-cell antigen receptor (TCR) and the costimulatory molecule CD28 that guide the access and retention of specific T cells into antigen-rich non-lymphoid tissue [1, 2]. This provides a new avenue for as yet unexplored therapeutic strategies targeting the inhibition of PI3K, with a view to altering T-cell migration at various stages of the immune response. This article will consider the known signaling events involved in T-cell migration during distinct modes of T-cell trafficking in response to either chemokines or the TCR and CD28.

Class 1 PI3Ks: An Overview

The class 1 PI3Ks are composed of a regulatory subunit and a tightly associated catalytic subunit. The class 1A enzymes are represented by five regulatory subunits encoded by three genes: *PIK3r1* encodes p85α and its alternative transcripts p55α and p50α. *PIK3r2* encodes p85β and *PIK3r3* encodes p55γ. The three class 1 catalytic isoforms p110α, p110β and p110δ pair with one of these regulatory subunits which are responsible for recruitment of the complex to the plasma membrane upon receptor ligation. Class 1A isoforms are activated downstream of immune cell receptors including the TCR, BCR, costimulatory molecules and cytokine receptors that are phosphorylated by tyrosine kinases upon cognate stimulus [3, 4]. The class 1B catalytic isoform p110γ pairs with either the regulatory subunits p84/p87 or p101 [5, 6] and is activated by G-protein βγ subunits and signals downstream of G-protein-coupled receptors (GPCRs). It is becoming increasingly apparent however that some GPCRs including chemokine receptors activate class IA PI3Ks, most notably p110β [7–9].

Expression of p110δ and PI3Kγ is largely restricted to leukocytes and therefore represent promising targets for selective inhibition of PI3K-mediated signaling pathways involved in inflammatory and autoimmune diseases. Mice in which the genes encoding p110δ or p110γ have been either ablated or altered to encode kinase-inactive versions, are viable, fertile and apparently healthy [7]. However, when their immune system is challenged, they exhibit severely altered phenotypes demonstrating that p110γ and p110δ have non-redundant functions in mast cells, neutrophils, dendritic cells, B and T cells, and that the activities of these isoforms in immune cells are crucial during the onset, progression and maintenance of chronic inflammatory diseases [7, 10]. Importantly, there is growing evidence that p110γ and p110δ act in partnership to regulate immune cell signaling and function [10]. Indeed it is interesting to note that mice deficient in both p110γ and p110δ (unlike mice deficient in single isoforms) display severe impairment of thymocyte development and profound T-cell lymphopenia as well as T-cell and eosinophil infiltration of mucosal organs, elevated IgE levels, and a skewing toward Th2 immune responses [11, 12]. However, the serious immune developmental defects observed in the p110γδ null mice prevent serious dissection of the selective roles of these p110 subunits in post-thymic responses.

A Role for PI3K in Cell Migration: The Story So Far

The major products of class I PI3Ks are 3′-phosphoinositides, most notably phosphatidylinositol 3,4,5-trisphosphate ($PI(3,4,5)P_3$) which functions as a key signaling molecule. The effects of $PI(3,4,5)P_3$ are counteracted by the lipid phosphatases PTEN and SHIP, which convert this lipid to $PI(4,5)P_2$ and $PI(3,4)P_2$ respectively [13]. $PI(3,4,5)P_3$ has important biological functions that rely on interaction with effector proteins containing lipid-binding domains such as pleckstrin homology domains [14, 15]. Several guanine nucleotide exchange factors, particular those with specificity for Rac, are regulated by 3′-phosphoinositides, while PI3K activity can itself be modulated by Rho GTPases [16–18]. Moreover, another well-characterized downstream PI3K effector, the Ser/Thr kinase Akt, has been implicated in F-actin polymerization and myosin assembly [19–22]. Accordingly, it has been proposed that PI3K sits centrally in an evolutionary conserved cell navigational mechanism, contributing to several aspects of the migratory machinery including gradient sensing, signal amplification, actin reorganization and hence cell motility [23–26]. Around 2002, several studies in neutrophils and *Dictyostelium* led to the notion that $PI(3,4,5)P_3$-dependent signals were part of a compass mechanism, sensing and responding to extracellular gradients of chemoattractants [16, 27–29]. First, use of biosensors composed of fluorescent proteins fused to PH domains that selectively bind $PI(3,4,5)P_3$ revealed that this lipid becomes highly polarized to the leading edge in amoebae and neutrophil-like cell lines [16, 27–29]. Second, PI3K inhibitors or genetic loss of PI3Ks reduced chemotactic responses of neutrophils and amoebae in both in vitro and in vivo migration assays [27, 28, 30–32]. Recent findings however have forced a re-evaluation of the role played by PI3Ks in cell navigational mechanisms. For example, some experiments examining the effects of either genetic loss of PI3Ks or selective PI3K inhibitors on the chemotactic efficiency of both neutrophils and *Dictyostelium* amoebae actually revealed no specific deficiencies [33, 34]. Furthermore, the $PI(3,4,5)P_3$ polarization to the leading edge of migrating cells was initially thought to be facilitated by the exclusion of PTEN from the leading edge and localization to the trailing edge of the migrating cell [27, 28, 35]. Evidence from neutrophils reveals that in fact, SHIP rather than PTEN, provides a critical role in the polarization and motility of these cells [36, 37]. Lastly, experiments examining the effects of either genetic loss of PI3Ks or selective PI3K inhibitors on the chemotactic efficiency of both neutrophils and *Dictyostelium* amoebae revealed no specific deficiencies [32, 34]. Remarkably, genetic loss of PI3Kγ or selective PI3K inhibitors actually reduced neutrophil chemokinetic cell responses [32]. This overall reduction in velocity and/or motility might explain some of the previously reported reductions in migration observed with pharmacological or gene targeting of PI3Kγ, rather than an impaired ability to move toward a chemoattractant gradient per se.

Evidence for PI3K-Dependent and -Independent T-Lymphocyte Directional Migration

Cell polarization, whereby the molecular processes at the front (leading edge) and the back (uropod) of a moving cell are different, is a prerequisite for efficient migration. It is well established in other systems that the small GTPases Rho, Rac and cdc42, have key roles in regulating cell polarity and morphology of migrating cells through effects on the actin cytoskeleton and actomyosin contraction and involves cross-talk with other signaling elements such as those provided by PI3K [38–40]. Indeed, Rho-dependent signaling is a key component of T-cell migration and adhesion in response to several chemokines in mature T cells and thymocytes, while polarization and migration of T-lymphocytes requires rapid Rac-driven formation of F-actin at the leading edge [2]. In this regard, several Rac guanine nucleotide exchange factors (GEFs) including DOCK2, Tiam1 and Vav have been implicated in T-lymphocyte adhesion and migratory events [2]. The extent of the involvement of PI3K-dependent signaling however has been less clearly defined and will be considered in detail below.

Activation of PI3K is a robust signaling event elicited by most homeostatic and inflammatory chemokine receptors expressed on T-lymphocytes [9, 41]. Chemokine interaction with GPCRs on lymphocytes in response to either homeostatic or inflammatory chemokines has been shown to depend predominantly on Gαi proteins [42]. This led to the assumption that these chemokines receptors are coupled to the βγ-dependent p110γ isoform. This is indeed the case, although several chemokine receptors can activate other PI3K isoforms [9, 41, 43, 44]. Early studies revealed that chemokine-stimulated migration of leukemic T-cell lines and primary T cells across synthetic membranes on transwell permeable supports in the absence of endothelial cells is abrogated by pan-isoform PI3K inhibitors [9, 41]. Closer inspection of the contribution of individual PI3K isoforms to T-lymphocyte migratory responses to chemokines utilized newly available genetic and pharmacological approaches. This revealed that the in vitro migration of p110γ-deficient CD4+ and CD8+ T cells to CCL19 and CCL21 and CXCL12 is significantly decreased compared to cells from wild-type mice [45]. Likewise, p110γ-targeting inhibitors (but not inhibitors directed toward the α, β or γ isoforms) inhibited migration of freshly isolated human peripheral blood T cells [44]. Remarkably, ex vivo maintenance and activation/differentiation of these human T cells leads to the migratory response acquiring a resistance to PI3K inhibitors [44, 46], indicating that the activation status of the cell helps determine whether PI3K is required for migratory responses to chemoattractants. Furthermore, PI3K inhibitors have little effect on T-cell migration in assays that better reflect physiological conditions. For example, T-lymphocyte arrest and adhesion to high endothelial venules in exteriorized Peyer's patches [47] or on transendothelial migration in laminar flow chambers [48] in response to either CXCR4 and/or CCR7 ligation is unaffected by PI3K inhibitors.

Other lines of evidence also cast further doubt as to whether the model for PI3K/PTEN polarization in neutrophils can be applied to T-lymphocytes. For example, many studies have been performed in the Jurkat leukemic T-cell line. These cells polarize and migrate normally in response to several chemokines acting on pertussis toxin-sensitive Gαi-coupled receptors, despite the fact that they are deficient in both PTEN and SHIP protein expression [9, 49, 50]. In fact, reconstitution of PTEN expression in Jurkat cells downregulated CXCL12-stimulated cell migration indicating a negative regulatory role for PTEN in T-cell migration [51]. Introduction of a constitutively active SHIP mutant into leukemic cell lines normally deficient in SHIP abrogates CXCL12- mediated migration [52]. This was somewhat surprising given the reported role of SHIP in neutrophil polarization [36]. However, this effect probably reflects that this construct is expressed widely throughout the plasma membrane and disrupts polarized accumulation of $PI(3,4,5)P_3$ at the leading edge.

Role of PI3K in T-Lymphocyte Homing and Migration in vivo: Lessons from Gene-Targeted from Mice

The use of genetically targeted mice in conjunction with in vivo models of homing of T cells to peripheral lymphoid nodes or effector T cells to sites of inflammation/antigen challenge has helped refine our understanding of the role of PI3K in T-cell migration. Analysis of mice lacking either the Rac-specific GEF DOCK2 and p110γ (alone or in combination), revealed that while DOCK-2 is the predominant molecule required for T-cell migration, p110γ can sustain a modest residual migratory response. Hence, optimum T-lymphocyte migration in vivo is dependent on expression of both DOCK2 and p110γ [53]. Importantly, a more recent study reported that p110γ is dispensable for constitutive migration of naive CD8 T cells, but plays a central role in the migration of effector CD8 T cells into inflammatory sites [54]. The reason for this discrepancy with earlier studies may reflect differences in the T-cell populations analyzed (e.g. bulk vs. CD8+ T cells). Although $p110\gamma^{-/-}$ T cells exhibit modest defects in migration in vitro and in vivo, it is notable that pan-isoform PI3K inhibitors such as wortmannin or Ly294002 effectively block in vitro and in vivo migration of naive T cells [44, 45]. This may simply reflect off-target effects of these compounds or the involvement of other PI3K isoforms in cell migration. Certainly, recent evidence has identified p110δ as being required for antigen-driven T-cell localization as discussed later [55], and is the dominant PI3K isoform in B-cell homing [45]. Interestingly, analysis of neutrophil migration in vivo revealed that in fact, while p110γ is important in early chemokine-induced emigration, p110δ replaces and maintains the delayed chemokine-induced neutrophil recruitment into inflamed tissues [56, 57]. Whether these isoforms fulfill a similar role during T-cell migration in vivo remains to be established. It is noteworthy however that recent work on natural killer (NK) cells (which are specialized lymphocytes linked to the innate immune response), has revealed differential roles for p110γ and p110δ in NK cell trafficking in health and disease. Thus, both p110γ and p110δ are necessary for NK cell migration to inflamed tissues and the uterus during early pregnancy in vivo and chemotaxis to CXCL12 and CCL3 in vitro. Remarkably, p110δ alone was required for NK cell distribution in steady state as well as for trafficking to lymphomas and for chemotaxis to S1P and CXCL10 in vitro [58].

Role of PI3K in Interstitial T-Lymphocyte Motility

T and B cells move vigorously within their specific microenvironments within SLO following apparently random migration pathways [59, 60]. This interstitial lymphocyte migration is integrin-independent, but instead is mediated by actin flow along the confining extracellular matrix scaffold structure, shape change and squeezing [61, 62]. It is likely that such random movement serves to increase dendritic cell screening efficiency, hence accelerating immune response initiation. Basal motility of T cells requires CCL19 and CCL21 (CCR7 agonists) that are abundant throughout the T-cell zone, together with adhesion ligands present on stromal cells [63, 64]. Multiphoton and conventional epifluorescence microscopy studies have explored whether PI3K is involved in regulating basal interstitial T-lymphocyte migration/motility within intact lymphoid tissue in vivo. Despite evidence of p110γ contributing to T-cell homing to lymphoid tissue and migration [53], there was no effect on the dynamic movements of p110γ-deficient T cells or the pan-PI3K isoform inhibitor wortmannin, compared to wild-type controls inside the T-cell area [63, 64]. Interestingly, another group using multiphoton microscopy in conjunction with wortmannin revealed a modest reduction of T- and B-cell velocities compared to untreated controls [65]. Complimentary gene targeting strategies in which class 1A function had been ablated by deletion of the *pik3r1* (p85α, p55α and p50α null) and *pik3r2* (p85β null) gene products showed a significant decrease in velocity and a marked loss of cell polarization. However, these experiments do not distinguish whether reduced motility results from impaired class IA PI3K signaling function or from loss of adaptor functions of the regulatory subunits independently of their role in activating the catalytic subunits. The reduced motility in wortmannin-treated cells supports at least some role for PI3K enzymatic subunits, but could also be due to inhibition of other PI3K subclasses or non-PI3K targets of wortmannin [10, 41, 44].

Antigen Recognition by the TCR and Costimulatory Receptors Influence T-Cell Trafficking

As well as being implicated in cell migration, PI3K has been a well-documented and robust signal elicited upon both TCR and CD28 ligation [66–69]. As reviewed extensively elsewhere, TCR- and costimulatory receptor-derived signals regulate both T-cell motility in vitro and trafficking in vivo, while the influence of Ag location on the localization, accumulation and retention of specific T cells is now well established from studies of animal models of autoimmunity and infection [1, 2]. Intravital microscopy has also revealed that antigen presentation by the endothelium selectively enhanced T-cell TEM without affecting rolling and adhesion [70]. Recent efforts have therefore focused on assessing the role of PI3K in antigen-mediated cell trafficking. To this end, the primary isoform coupled to TCR and CD28 is believed to be p110δ [67, 71, 72]. TCR-transgenic mice carrying an ovalbumin-specific T-cell receptor (OT-II) and a mutation in the cytoplasmic tail of CD28 that abrogates class I PI3K recruitment without leading to defects in clonal expansion (OT-II/CD28^{Y170F}) [73] were generated to allow discrimination of conventional costimulation-driven clonal expansion from the ability to infiltrate antigenic tissue. OT-II and OT-II/CD28^{Y170F} naive T cells proliferated equivalently following immunization with $OVA_{323-339}$ peptide. However, OT-II/CD28^{Y170F} CD8+ memory T cells failed to localize to target tissue upon antigen challenge. These findings were reinforced by subsequent studies using T cells from mice expressing a catalytically inactive p110δ isoform that revealed an essential role for p110δ in TCR-dependent localization of both CD4+ and CD8+ T cells in a male antigen-specific transplantation model [55]. Interestingly, and in support of previous findings [45], there was no defect in the p110δ mutant mice of either normal constitutive trafficking or migratory response to non-specific chemokine agonists [55]. Defects in TCR-induced T-cell proliferation and signaling have been reported in p110γ-deficient T cells [74]. Hence it is important to highlight recent work using the OT-II transgenic TCR model in mice lacking p110γ reveals no defect in TCR signaling or proliferation in response to antigen, yet their ability to traffick to peripheral inflammatory sites in vivo was severely impaired [75]. This was interpreted as a consequence of the inability of p110$\gamma^{-/-}$ cells to migrate toward inflammatory chemokines that prevented migration to inflammatory sites. Certainly, signals provided by chemokines permit the full crossing of the endothelial barrier and the completion of antigen-dependent TEM [76] and it seems likely that there should be cooperation between TCR and chemokine-mediated signals in the regulation of T-cell migration. Indeed, there is evidence of direct cross-talk between TCR- and chemokine receptor-mediated signaling. In this regard, both ZAP-70, SLP-76, Tec kinases (key elements in TCR signaling), have been implicated in CXCR4 signal transduction in human T cells [77, 78]. Moreover, CXCL12 (the ligand for CXCR4) stimulates a physical association between CXCR4 and the TCR and utilizes the ZAP-70-binding immunoreceptor tyrosine-based activation motifs of the TCR for signal transduction [79, 80]. So, it seems that p110δ and p110γ are likely to play complimentary roles in the migration of effector cells out of vessels and into tissues, with their relative contribution shaped by the nature and timing of receptor activation.

PI3K Influences Effector T-Cell Migration at Transcriptional and Post-Translational Levels

It is becoming clear that an additional route by which PI3K can influence cell migration is via the regulation of transcription factors that regulate cell quiescence and expression of homing receptors on T cells. Hence, genetic and pharmacological studies have revealed p110δ plays an essential role in the events that lead to proteolytic shedding and reduced transcription of CD62L, CCR7 and S1P receptor-1 (S1P1) [81]. These surface proteins play an essential role in homing of CD8– cells to secondary lymphoid tissues and prevent their egress to sites of peripheral inflammation. A critical event in this pathway is downregulation of the transcription factor Kruppel-like factor-2 (KLF2), which promotes L-selectin, CCR7 and $S1P_1$ transcription [82, 83]. Inhibition of individual components of the PI3K-dependent pathway prevented the loss of L-selectin and CCR7. In particular, the authors show that the mTOR inhibitor rapamycin prevents loss of L-selectin and CCR7 and further demonstrate that rapamycin-treated CD8+ cells preferentially home to lymphoid tissues rather than peripheral sites. The ability of rapamycin to contain activated effector cells in SLO could result in the destruction of antigen-primed dendritic cells and termination of immune response [84] and prevent immune destruction of target cells in the periphery, thus providing a previously undiscovered mechanism of action of rapamycin as an immune suppressant drug. Another study revealed that the transcription factor FOXO1 exerts a steady-state control on L-selectin expression in resting T cells and this is opposed by PI3K [85]. Central to the regulation of FOXOs by PI3K is their phosphorylation by the PI3K effector protein kinase B/Akt. As a consequence, FOXO molecules are excluded from the nucleus and their transcriptional activities switched off in the activated cell [85]. These studies therefore highlight the importance of transcriptional events regulated by PI3K will likely influence antigen-dependent trafficking T cells to tissue and this will be an important avenue for future research.

Conclusions

Several therapeutic strategies have been explored to prevent leukocyte migration, including blockade of adhesion molecules, chemokine receptors and signaling events such as those mediated by PI3K [86]. The diverse milieu of chemokines, adhesion ligands and stromal cell architecture in different regions within lymphoid organs and peripheral tissues as well as the different state of activation of individual cells, will influence the expression of antigen, costimulatory and chemoattractant receptors. In turn, these will determine the relative contribution of individual PI3K isoforms versus other biochemical signals to migratory machinery (fig. 1). The considerable scope for plasticity of PI3K isoform involvement in lymphocyte migration is illustrated by the way that p110γ and p110δ (as in neutrophils) are both equally required for PI(3,4,5)

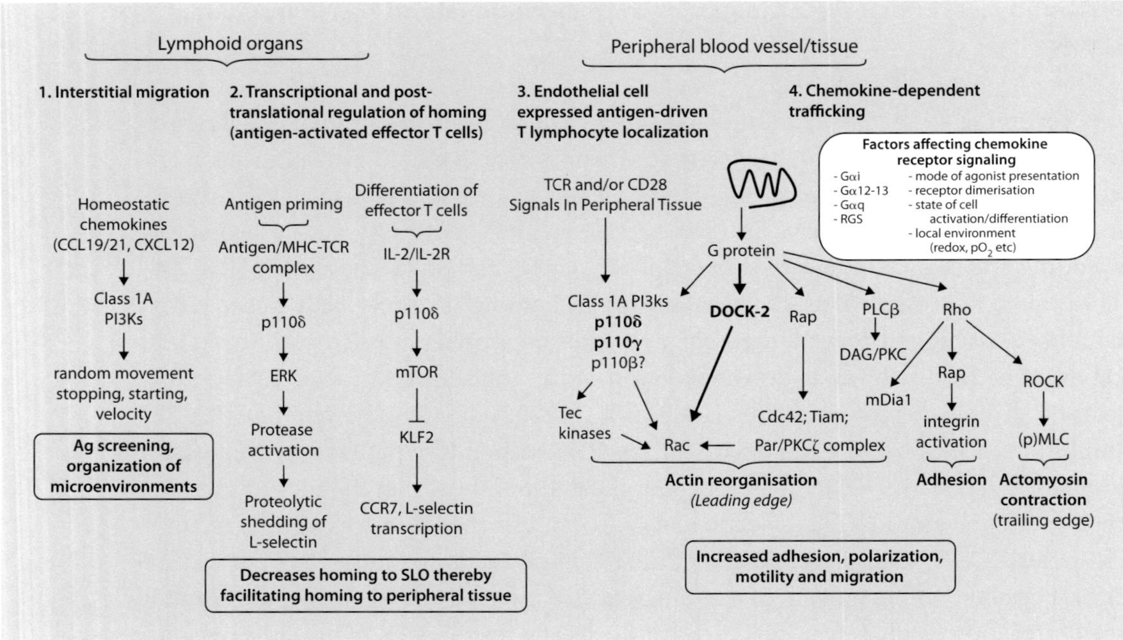

Fig. 1. PI3K influences chemokine-stimulated T-lymphocyte migration and motility. Class 1 PI3K isoforms influence T-lymphocyte migration at several levels in lymphoid and peripheral tissue: (1) Class 1A PI3K isoforms influence random interstitial cell migration events in lymphoid organs that underpins Ag screening and tissue architecture, although the identity of isoforms involved is unknown. (2) Activation of p110δ by antigen-engaged TCR and IL-2 receptor signaling during T-cell growth and differentiation mediates downregulation of L-selectin and CCR7 expression by either proteolytic shedding and/or decreased transcription. (3) Activation of p110δ by the TCR and costimulatory receptors upon recognition of antigen and B7 family molecules displayed by the endothelium contributes to antigen-driven T-lymphocyte localization. (4) DOCK-2 is the predominant signal that leads to Rac activation and initial actin reorganization (as denoted by larger text/arrows), although there does appear to be a significant contribution provided p110γ depending on the context of cell migration. The signaling pathways linked to adhesion and formation of trailing edge are also summarized. The precise balance of signaling via p110γ (or other class 1 PI3Ks) versus DOCK-2 and other pathways leading to actin reorganization, cell polarization and adhesion will be shaped by the type of G-protein subunits to which individual receptors are coupled and other factors that are indicated. DAG = Diacylglycerol; ERK1 = extracellular signal-regulated kinase; KLF-2 = Kruppel-like factor-2; mDia = mammalian diaphanous-related formin; PKC = protein kinase-C; PLC = phospholipase C; (p)-MLC = phosphomyosin light chain; mTOR = mammalian target of rapamycin; RGS = regulator of G-protein signaling (RGS); Tiam = T-cell lymphoma invasion and metastasis.

P_3 production in response to CXCR4 in NK cells, whereas in T cells, the response to the same GPCR is exclusively p110γ-dependent. Moreover, p110δ is indispensible for NK migration during pregnancy, inflammation, steady-state recirculation and trafficking to lymphomas, while the requirement for p110γ was restricted to pregnancy and inflammation only [58]. One might imagine therefore heterogenous coupling of individual chemokine receptors to PI3K isoforms in different T-cell subsets at varying states of activation. The promiscuity of certain chemokine receptors for their cognate ligands provides further opportunity for plasticity of PI3K isoform coupling.

The differing dependence of individual chemokine receptors and antigen receptors on PI3K isoforms at different stages of lymphocyte activation and in different lymphocyte subsets makes it difficult to design a 'one-fits-all' drug to inhibit inflammatory recruitment of T-lymphocytes. Both genetic and pharmacological strategies have revealed that p110γ can make a contribution to the migration of naive T cells to lymph nodes while both p110γ and p110δ contribute to the migration of effector T cells as well as NK cells to sites of inflammation. Our recently acquired appreciation of the role of p110δ in regulating primed T-cell migration to antigenic sites provides an additional dimension to its potential as a pharmacological target to control of T-cell-mediated pathologies, including autoimmunity and transplantation. Hence, selective targeting of p110δ may avoid undesired T-cell-dependent inflammation by preventing antigen-dependent T-cell migration and subsequently cell-cell interactions without inducing overt immune suppression.

Acknowledgement

S.G.W. is the recipient of a Royal Society Industrial Fellowship.

References

1 Marelli-Berg FM, Cannella L, Dazzi F, Mirenda V: The highway code of T cell trafficking. J Pathol 2008;214:179–189.
2 Ward SG, Marelli-Berg FM: Mechanisms of chemokine and antigen-dependent T-lymphocyte navigation. Biochem J 2009;418:13–427.
3 Deane JA, Fruman DA: Phosphoinositide 3-kinase: diverse roles in immune cell activation. Annu Rev Immunol 2004;22:563–598.
4 Ward SG, Cantrell DA: Phosphoinositide 3-kinases in T-lymphocyte activation. Curr Opin Immunol 2001;13:332–338.
5 Suire S, Coadwell J, Ferguson GJ, Davidson K, Hawkins P, Stephens L: p84, a new Gβγ-activated regulatory subunit of the type IB phosphoinositide 3-kinase p110γ. Curr Biol 2005;15:566–570.
6 Voigt P, Dorner MB, Schaefer M: Characterization of p87PIKAP, a novel regulatory subunit of phosphoinositide 3-kinase-γ that is highly expressed in heart and interacts with PDE3B. J Biol Chem 2006;281:9977–9986.
7 Vanhaesebroeck B, Ali K, Bilancio A, Geering B, Foukas LC: Signalling by PI3K isoforms: insights from gene-targeted mice. Trends Biochem Sci 2005;30:194–204.
8 Guillermet-Guibert J, Bjorklof K, Salpekar A, Gonella C, Ramadani F, Bilancio A, Meek S, Smith AJ, Okkenhaug K, Vanhaesebroeck B: The p110β isoform of phosphoinositide 3-kinase signals downstream of G-protein-coupled receptors and is functionally redundant with p110γ. Proc Natl Acad Sci USA 2008;105:8292–8297.
9 Ward SG: Do phosphoinositide 3-kinases direct lymphocyte navigation? Trends Immunol 2004;25:67–74.
10 Crabbe T, Welham MJ Ward SG: The PI3K inhibitor arsenal: choose your weapon! Trends Biochem Sci 2007;32:450–456.
11 Webb LM, Vigorito E, Wymann MP, Hirsch E, Turner M: T cell development requires the combined activities of the p110γ and p110δ catalytic isoforms of phosphatidylinositol 3-kinase. J Immunol 2005;175:2783–2787.
12 Ji H, Rintelen F, Waltzinger C, Bertschy Meier D, Bilancio A, Pearce W, Hirsch E, Wymann MP, Ruckle T, Camps M, Vanhaesebroeck B, Okkenhaug K, Rommel C: Inactivation of PI3Kγ and PI3Kδ distorts T-cell development and causes multiple organ inflammation. Blood 2007;110:2940–2947.

13 Harris SJ, Parry RV, Westwick J, Ward SG: Phosphoinositide lipid phosphatases: natural regulators of phosphoinositide 3-kinase signaling in T-lymphocytes. J Biol Chem 2008;283:2465–2469.
14 Ellson CD, Andrews S, Stephens LR, Hawkins PT: The PX domain: a new phosphoinositide-binding module. J Cell Sci 2002;115:1099–1105.
15 Lemmon MA, Ferguson KM: Signal-dependent membrane targeting by pleckstrin homology domains. Biochem J 2000;350:1–18.
16 Weiner OD, Neilsen PO, Prestwich GD, Kirschner MW, Cantley LC, Bourne HR: A $PtdInsP_3$- and Rho GTPase-mediated positive feedback loop regulates neutrophil polarity. Nat Cell Biol 2002;4:509–513.
17 Andrews S, Stephens LR, Hawkins PT: PI3K class IB pathway in neutrophils. Sci STKE 2007;2007:cm3.
18 Andrews S, Stephens LR, Hawkins PT: PI3K class IB pathway. Sci STKE 2007;2007:cm2.
19 Stambolic V, Woodgett JR: Functional distinctions of protein kinase B/Akt isoforms defined by their influence on cell migration. Trends Cell Biol 2006; 16:461–466.
20 Kolsch V, Charest PG, Firtel RA: The regulation of cell motility and chemotaxis by phospholipid signaling. J Cell Sci 2008;121:551–559.
21 Li J, Ballif BA, Powelka AM, Dai J, Gygi SP, Hsu VW: Phosphorylation of ACAP1 by Akt regulates the stimulation-dependent recycling of integrin-β_1 to control cell migration. Dev Cell 2005;9:663–673.
22 Enomoto A, Murakami H, Asai N, Morone N, Watanabe T, Kawai K, Murakumo Y, Usukura J, Kaibuchi K, Takahashi M: Akt/PKB regulates actin organization and cell motility via Girdin/APE. Dev Cell 2005;9:389–402.
23 Ridley AJ, Schwartz MA, Burridge K, Firtel RA, Ginsberg MH, Borisy G, Parsons JT, Horwitz AR: Cell migration: integrating signals from front to back. Science 2003;302:1704–1709.
24 Merlot S, Firtel RA: Leading the way: Directional sensing through phosphatidylinositol 3-kinase and other signaling pathways. J Cell Sci 2003;116:3471–3478.
25 Hogg N, Laschinger M, Giles K, McDowall A: T-cell integrins: more than just sticking points. J Cell Sci 2003;116:4695–4705.
26 Stephens L, Milne L, Hawkins P: Moving towards a better understanding of chemotaxis. Curr Biol 2008; 18:R485–494.
27 Funamoto S, Meili R, Lee S, Parry L, Firtel RA: Spatial and temporal regulation of 3-phosphoinositides by PI 3-kinase and PTEN mediates chemotaxis. Cell 2002;109:611–623.
28 Iijima M, Devreotes P: Tumor suppressor PTEN mediates sensing of chemoattractant gradients. Cell 2002;109:599–610.
29 Wang F, Herzmark P, Weiner OD, Srinivasan S, Servant G, Bourne HR: Lipid products of PI3Ks maintain persistent cell polarity and directed motility in neutrophils. Nat Cell Biol 2002;4:513–518.
30 Hannigan M, Zhan L, Li Z, Wu D, Huang C: Neutrophils lacking phosphoinositide 3-kinase-γ show loss of directionality during *N*-formyl-Met-Leu-Phe-induced chemotaxis. Proc Natl Acad Sci USA 2002;99:3603–3608.
31 Hirsch E, Katanaev VL, Garlanda C, Azzolino O, Pirola L, Silengo L, Sozzani S, Mantovani A, Altruda F, Wymann MP: Central role for G-protein-coupled phosphoinositide 3-kinase-γ in inflammation. Science 2000;287:1049–1053.
32 Ferguson GJ, Milne L, Kulkarni S, Sasaki T, Walker S, Andrews S, Crabbe T, Finan P, Jones G, Jackson S, Camps M, Rommel C, Wymann M, Hirsch E, Hawkins P, Stephens L: PI3Kγ has an important context-dependent role in neutrophil chemokinesis. Nat Cell Biol 2007;9:86–91.
33 Hoeller O, Kay RR: Chemotaxis in the absence of PIP3 gradients. Curr Biol 2007;17:813–817.
34 Andrew N, Insall RH: Chemotaxis in shallow gradients is mediated independently of PtdIns 3-kinase by biased choices between random protrusions. Nat Cell Biol 2007;9:193–200.
35 Li Z, Hannigan M, Mo Z, Liu B, Lu W, Wu Y, Smrcka AV, Wu G, Li L, Liu M, Huang CK, Wu D: Directional sensing requires G β γ-mediated PAK1 and PIX α-dependent activation of Cdc42. Cell 2003;114:215–227.
36 Nishio M, Watanabe K, Sasaki J, Taya C, Takasuga S, Iizuka R, Balla T, Yamazaki M, Watanabe H, Itoh R, Kuroda S, Horie Y, Forster I, Mak TW, Yonekawa H, Penninger JM, Kanaho Y, Suzuki A, Sasaki T: Control of cell polarity and motility by the $PtdIns(3,4,5)P_3$ phosphatase SHIP1. Nat Cell Biol 2007;9:36–44.
37 Subramanian KK, Jia Y, Zhu D, Simms BT, Jo H, Hattori H, You J, Mizgerd JP, Luo HR: Tumor suppressor PTEN is a physiologic suppressor of chemoattractant-mediated neutrophil functions. Blood 2007;109:4028–4037.
38 Etienne-Manneville S, Hall A: Rho GTPases in cell biology. Nature 2002;420:629–635.
39 Charest PG, Firtel RA: Big roles for small GTPases in the control of directed cell movement. Biochem J 2007;401:377–390.
40 Pollard TD, Borisy GG: Cellular motility driven by assembly and disassembly of actin filaments. Cell 2003;112:453–465.
41 Ward SG: T-lymphocytes on the move: chemokines, PI 3-kinase and beyond. Trends Immunol 2006;27: 80–87.

42 Han SB, Moratz C, Huang NN, Kelsall B, Cho H, Shi CS, Schwartz O, Kehrl JH: Rgs1 and Gnai2 regulate the entrance of B lymphocytes into lymph nodes and B-cell motility within lymph node follicles. Immunity 2005;22:343–354.

43 Sotsios Y, Ward SG: Phosphoinositide 3-kinase: a key biochemical signal for cell migration in response to chemokines. Immunol Rev 2000;177:217–235.

44 Smith LD, Hickman ES, Parry RV, Westwick J, Ward SG: PI3Kγ is the dominant isoform involved in migratory responses of human T-lymphocytes: effects of ex vivo maintenance and limitations of non-viral delivery of siRNA. Cell Signal 2007;19:2528–2539.

45 Reif K, Okkenhaug K, Sasaki T, Penninger JM, Vanhaesebroeck B, Cyster JG: Differential roles for phosphoinositide 3-kinases, p110γ and p110δ, in lymphocyte chemotaxis and homing. J Immunol 2004;173:2236–2240.

46 Smit MJ, Verdijk P, van der Raaij-Helmer EM, Navis M, Hensbergen PJ, Leurs R, Tensen CP: CXCR3-mediated chemotaxis of human T cells is regulated by a Gi- and phospholipase C-dependent pathway and not via activation of MEK/p44/p42 MAPK nor Akt/PI-3 kinase. Blood 2003;102:1959–1965.

47 Constantin G, Majeed M, Giagulli C, Piccio L, Kim JY, Butcher EC, Laudanna C: Chemokines trigger immediate β_2 integrin affinity and mobility changes: differential regulation and roles in lymphocyte arrest under flow. Immunity 2000;13:759–769.

48 Cinamon G, Shinder V, Alon R: Shear forces promote lymphocyte migration across vascular endothelium bearing apical chemokines. Nat Immunol 2001;2:515–521.

49 Astoul E, Edmunds C, Cantrell DA, Ward SG: PI 3-K and T-cell activation: limitations of T-leukemic cell lines as signaling models. Trends Immunol 2001;22:490–496.

50 Freeburn RW, Wright KL, Burgess SJ, Astoul E, Cantrell DA, Ward SG: Evidence that SHIP-1 contributes to phosphatidylinositol 3,4,5-trisphosphate metabolism in T lymphocytes and can regulate novel phosphoinositide 3-kinase effectors. J Immunol 2002;169:5441–5450.

51 Gao P, Wange RL, Zhang N, Oppenheim JJ, Howard OM: Negative regulation of CXCR4-mediated chemotaxis by the lipid phosphatase activity of tumor suppressor PTEN. Blood 2005;106:2619–2626.

52 Wain CM, Westwick J, Ward SG: Heterologous regulation of chemokine receptor signaling by the lipid phosphatase SHIP in lymphocytes. Cell Signal 2005;17:1194–1202.

53 Nombela-Arrieta C, Lacalle RA, Montoya MC, Kunisaki Y, Megias D, Marques M, Carrera AC, Manes S, Fukui Y, Martinez AC, Stein JV: Differential requirements for DOCK2 and phosphoinositide-3-kinase-γ during T and B lymphocyte homing. Immunity 2004;21:429–441.

54 Martin AL, Schwartz MD, Jameson SC, Shimizu Y: Selective regulation of CD8 effector T cell migration by the p110γ isoform of phosphatidylinositol 3-kinase. J Immunol 2008;180:2081–2088.

55 Jarmin SJ, David R, Ma L, Chai JG, Dewchand H, Takesono A, Ridley AJ, Okkenhaug K, Marelli-Berg FM: T-cell receptor-induced phosphoinositide-3-kinase p110δ activity is required for T-cell localization to antigenic tissue in mice. J Clin Invest 2008;118:1154–1164.

56 Puri KD, Doggett TA, Huang CY, Douangpanya J, Hayflick JS, Turner M, Penninger J, Diacovo TG: The role of endothelial PI3Kγ activity in neutrophil trafficking. Blood 2005;106:150–157.

57 Liu L, Puri KD, Penninger JM, Kubes P: Leukocyte PI3Kγ and PI3Kδ have temporally distinct roles for leukocyte recruitment in vivo. Blood 2007;110:1191–1198.

58 Saudemont A, Garcon F, Yadi H, Roche-Molina M, Kim N, Segonds-Pichon A, Martin-Fontecha A, Okkenhaug K, Colucci F: p110γ and p110δ isoforms of phosphoinositide 3-kinase differentially regulate natural killer cell migration in health and disease. Proc Natl Acad Sci USA 2009;106:5795–5800.

59 Cahalan MD, Parker I: Imaging the choreography of lymphocyte trafficking and the immune response. Curr Opin Immunol 2006;18:476–482.

60 Cahalan MD, Parker I: Choreography of cell motility and interaction dynamics imaged by two-photon microscopy in lymphoid organs. Annu Rev Immunol 2008;26:585–626.

61 Lammermann T, Bader BL, Monkley SJ, Worbs T, Wedlich-Soldner R, Hirsch K, Keller M, Forster R, Critchley DR, Fassler R, Sixt M: Rapid leukocyte migration by integrin-independent flowing and squeezing. Nature 2008;453:51–55.

62 Friedl P, Weigelin B: Interstitial leukocyte migration and immune function. Nat Immunol 2008;9:960–969.

63 Asperti-Boursin F, Real E, Bismuth G, Trautmann A, Donnadieu E: CCR7 ligands control basal T cell motility within lymph node slices in a phosphoinositide 3-kinase-independent manner. J Exp Med 2007;204:1167–1179.

64 Worbs T, Mempel TR, Bolter J, von Andrian UH, Forster R: CCR7 ligands stimulate the intranodal motility of T-lymphocytes in vivo. J Exp Med 2007;204:489–495.

65 Matheu, MP, Deane JA, Parker I, Fruman DA, Cahalan MD: Class IA phosphoinositide 3-kinase modulates basal lymphocyte motility in the lymph node. J Immunol 2007;179:2261–2269.
66 Okkenhaug K, Ali K, Vanhaesebroeck B: Antigen receptor signalling: a distinctive role for the p110δ isoform of PI3K. Trends Immunol 2007;28:80–87.
67 Okkenhaug K, Bilancio A, Farjot G, Priddle H, Sancho S, Peskett E, Pearce W, Meek SE, Salpekar A, Waterfield MD, Smith AJ, Vanhaesebroeck B: Impaired B and T cell antigen receptor signaling in p110δ PI 3-kinase mutant mice. Science 2002;297: 1031–1034.
68 Ward SG: CD28: a signalling perspective. Biochem J 1996;318:361–77.
69 Ward SG, June CH, Olive D: PI 3-kinase: a pivotal pathway in T-cell activation? Immunol Today 1996; 17:187–197.
70 Marelli-Berg FM, James MJ, Dangerfield J, Dyson J, Millrain M, Scott D, Simpson E, Nourshargh S, Lechler RI: Cognate recognition of the endothelium induces HY-specific CD8+ T-lymphocyte transendothelial migration (diapedesis) in vivo. Blood 2004; 103:3111–3116.
71 Okkenhaug K, Patton DT, Bilancio A, Garcon F, Rowan WC, Vanhaesebroeck B: The p110δ isoform of phosphoinositide 3-kinase controls clonal expansion and differentiation of Th cells. J Immunol 2006; 177:5122–5128.
72 Okkenhaug K, Vanhaesebroeck B: PI3K in lymphocyte development, differentiation and activation. Nat Rev Immunol 2003;3:317–330.
73 Okkenhaug K, Wu L, Garza KM, La Rose J, Khoo W, Odermatt B, Mak TW, Ohashi PS, Rottapel R: A point mutation in CD28 distinguishes proliferative signals from survival signals. Nat Immunol 2001;2: 325–332.
74 Alcazar I, Marques M, Kumar A, Hirsch E, Wymann M, Carrera AC, Barber DF: Phosphoinositide 3-kinase-γ participates in T cell receptor-induced T cell activation. J Exp Med 2007;204:2977–2987.
75 Thomas MS, Mitchell JS, Denucci CC, Martin AL, Shimizu Y: The p110γ isoform of phosphatidylinositol 3-kinase regulates migration of effector CD4 T lymphocytes into peripheral inflammatory sites. J Leukoc Biol 2008;84:814–823.
76 Manes TD, Pober JS: Antigen presentation by human microvascular endothelial cells triggers ICAM-1-dependent transendothelial protrusion by, and fractalkine-dependent transendothelial migration of, effector memory CD4+ T cells. J Immunol 2008;180:8386–8392.
77 Kremer KN, Humphreys TD, Kumar A, Qian NX, Hedin KE: Distinct role of ZAP-70 and Src homology-2 domain-containing leukocyte protein of 76 kDa in the prolonged activation of extracellular signal-regulated protein kinase by the stromal cell-derived factor-1α/CXCL12 chemokine. J Immunol 2003;171:360–367.
78 Ticchioni M, Charvet C, Noraz N, Lamy L, Steinberg M, Bernard A, Deckert M: Signaling through ZAP-70 is required for CXCL12-mediated T-cell transendothelial migration. Blood 2002;99:3111–3118.
79 Alsayed Y, Ngo H, Runnels J, Leleu X, Singha UK, Pitsillides CM, Spencer JA, Kimlinger T, Ghobrial JM, Jia X, Lu G, Timm M, Kumar A, Cote D, Veilleux I, Hedin KE, Roodman GD, Witzig TE, Kung AL, Hideshima T, Anderson KC, Lin CP, Ghobrial IM: Mechanisms of regulation of CXCR4/SDF-1 (CXCL12)-dependent migration and homing in multiple myeloma. Blood 2007;109:2708–2717.
80 Kumar A, Humphreys TD, Kremer KN, Bramati PS, Bradfield L, Edgar CE, Hedin KE: CXCR4 physically associates with the T cell receptor to signal in T cells. Immunity 2006;25:213–224.
81 Sinclair LV, Finlay D, Feijoo C, Cornish GH, Gray A, Ager A, Okkenhaug K, Hagenbeek TJ, Spits H, Cantrell DA: Phosphatidylinositol-3-OH kinase and nutrient-sensing mTOR pathways control T-lymphocyte trafficking. Nat Immunol 2008;9:513–521.
82 Rollins BJ: Innocents abroad: regulating where naive T cells go. Nat Immunol 2008;9:233–235.
83 Sebzda E, Zou Z, Lee JS, Wang T, Kahn ML: Transcription factor KLF2 regulates the migration of naive T cells by restricting chemokine receptor expression patterns. Nat Immunol 2008;9:292–300.
84 Guarda G, Hons M, Soriano SF, Huang AY, Polley R, Martin-Fontecha A, Stein JV, Germain RN, Lanzavecchia A, Sallusto F: L-selectin-negative CCR7 effector and memory CD8+ T cells enter reactive lymph nodes and kill dendritic cells. Nat Immunol 2007;8:743–752.
85 Fabre S, Carrette F, Chen J, Lang V, Semichon M, Denoyelle C, Lazar V, Cagnard N, Dubart-Kupperschmitt A, Mangeney M, Fruman DA, Bismuth G: FOXO1 regulates L-selectin and a network of human T cell homing molecules downstream of phosphatidylinositol 3-kinase. J Immunol 2008;181:2980–2989.
86 Mackay CR: Moving targets: cell migration inhibitors as new anti-inflammatory therapies. Nat Immunol 2008;9:988–998.

Prof. Stephen Ward
Department of Pharmacy and Pharmacology, University of Bath
Claverton Down, Bath BA2 7AY (UK)
Tel. +44 225 323641, Fax +44 225 386114, E-Mail S.G.Ward@bath.ac.uk

Entschladen F, Zänker KS (eds): Cell Migration: Signalling and Mechanisms.
Transl Res Biomed. Basel, Karger, 2010, vol 2, pp 67–82

Migration of Functionally Specialized T-Helper Cells: T_{FH} Cells, Th17 Cells and FoxP3$^+$ T Cells

Chang H. Kim

Laboratory of Immunology and Hematopoiesis, Department of Comparative Pathobiology, Purdue Cancer Center, Purdue University, West Lafayette, Ind., USA

Abstract

The three subsets of T-helper cells, T_{FH} cells, Th17 cells and FoxP3$^+$ T cells, perform important functions in the immune system that are highly specialized and distinct from each other. T_{FH} cells positively regulate B-cell differentiation at multiple stages. Th17 cells mediate antibacterial and antifungal immune responses and are implicated in autoimmune diseases. FoxP3$^+$ T cells perform suppressive functions in regulation of immune responses and are important for promotion of tolerance. These T-helper cells are not only distinct in their function but also are different from each other in their migration ability. T_{FH} cells express CXCR5, the chemokine ligand of which is specifically expressed in B-cell follicles and stay within the secondary lymphoid tissues. Th17 cells highly express CCR6 and other memory/effector-type trafficking receptors to migrate to the intestine and inflamed tissues. FoxP3$^+$ T cells have both naive and memory-type trafficking receptors and effectively migrate to both lymphoid and non-lymphoid tissues to counterbalance all types of effector T cells. Additional trafficking receptors can further diversify the migration behavior of the T-helper cells for tissue-specific migration. I will review our recent understanding of the roles of migration and trafficking receptors in proper functioning of the specialized T-helper cell lineages.

Recently, the three T-cell subsets, Th17 cells, T_{FH} cells and FoxP3$^+$ T cells, have been getting a lot of attention. These T-helper cells have been actively studied in the last 5- to 10-year period. Th17 cells were named after their major cytokine product IL-17 [1]. T_{FH} cells were named after their tissue tropism for B-cell follicles [2, 3]. T_{FH} cells are heterogeneous: in addition to the T_{FH} cells outside of the follicles, there is a functionally mature T_{FH} subset in germinal centers which is commonly called 'GC-T cell' or 'GC T_{FH} cell' [4]. FoxP3$^+$ T cells are more frequently called T-regulatory cells or Tregs, but this name has been comprehensively used to refer to any T cells with suppressive functions [5]. Th17 cells were initially thought to function as inflammatory

T cells that induce autoimmune diseases in joints and the central nervous system (CNS) [6]. It is increasingly clear that Th17 cells are essential effector T cells that promote the immunity to bacteria and fungi [7]. T_{FH} cells help B cells with their differentiation into memory and plasma B cells [8]. FoxP3$^+$ T cells play roles very different from those of Th17 cells and T_{FH} cells. The goal of FoxP3$^+$ T cells is to dampen the function of many immune cell types including Th1, Th2, Th17 cells and T_{FH} cells. Immune responses are generally self-limited to prevent unwanted autoimmunity following elimination of pathogens, and FoxP3$^+$ T cells play critical roles in this regard [9]. I will review the process of generation, trafficking receptors, and migration of the three important T-helper cell subsets.

Migration and Trafficking Receptors of T Cells

Migration is a process critical for immune cells to perform their functions in the right tissues. This is particularly true for T cells, which are made in the thymus and emigrate to the blood system for recirculation. Recirculation is important for surveillance of antigens by naive T cells for induction of adaptive immunity and formation of immune memory. It is important also for dissemination, propagation, and reactivation of memory T cells. Recirculation of naive T cells occurs between the blood and secondary lymphoid tissues (SLT). Memory T cells can additionally patrol other peripheral tissues. T cells stop recirculating when they undergo activation in SLT. Sphingosin-1-phosphate (S1P) plays an important role in emigration of T cells from the thymus and lymphoid tissues. S1P has five receptors, S1P1 through S1P5. S1P acts as a chemoattractant for S1P1-expressing T cells [10]. The concentration of S1P is highest in the blood and lowest within thymus and SLT. The concentration in lymph is between those of blood and organs [11]. Therefore, there is a chemotactic gradient of S1P formed to induce emigration of lymphocytes to the lymph and then to blood [11]. S1P may act on endothelial cells to indirectly promote transendothelial emigration of lymphocytes [12]. When T cells undergo activation, S1P1 is downregulated to a level that it is no longer functional to respond to the S1P gradient. At the same time, activated T cells express adhesion molecules that keep the T cells in close contact to stromal cells and antigen-presenting cells [13]. When T cells complete their processes of activation and differentiation, they regain expression of S1P1 but downregulate adhesion molecules for emigration into the blood. FTY720 is a sphingosine analog derived from a fungal metabolite and induces prolonged downregulation of S1P1 on lymphocytes [14]. Therefore, FTY720 can block lymphocyte recirculation and is being tested as a new form of immunosuppressant drug. In a manner similar to the effect of FTY720, S1P1 null T cells cannot emigrate the thymus and, thus, are not able to recirculate [15, 16].

Naive T cells express high levels of CCR7, CXCR4, and CD62L and low levels of $\alpha_4\beta_7$ [5, 17, 18]. CD62L mediates the process of rolling of naive T cells on the high endothelial venules of peripheral lymph nodes, while $\alpha_4\beta_7$ or CD62L mediates the rolling on the endothelial cell surface of mesenteric lymph nodes [19]. CCR7 and CXCR4 mediate activation of integrins for firm adhesion and chemotaxis to the periarterial T zone of SLT. An important event occurring during antigen priming of naive T cells in SLT is the switch of the trafficking receptors from the naive type (CCR7, CXCR4, and CD62L) to the memory/effector type (CCR2, CCR4, CCR5, CCR6, CXCR3, CXCR5, CXCR6, P-selectin ligand, and E-selectin ligand) [20, 21]. While the naive-type receptors guide the T cells into SLT, the memory-type receptors guide T cells to inflamed tissues or other peripheral tissues such as the intestine. It has been well established that Th1 and Th2 cells differ from each other in expression of trafficking receptors and migrate to distinct effector tissues. Th1 cells express CCR5, CXCR3 and CXCR6, while Th2 cells express CRTH2, CCR3, and CCR4 [21, 22]. Also, Th17 cells and T_{FH} cells are different from each other in expression of trafficking receptors and migration [23–26]. In addition to the differences among the T-cell subsets, memory T cells that migrate to the intestine are different in expression of trafficking receptors from the T cells that migrate to the skin and other tissues [19, 27]. Gut homing cells generally require $\alpha_4\beta_7$ for the migration to the intestine [28–30], and those that migrate to the small intestine additionally need CCR9 [31–35]. However, some CCR9-deficient cells still can migrate to the small intestine, suggesting the role of alternative receptors in migration to the small intestine [31]. T cells that migrate to inflamed skin use E/P-selectin ligands, CCR4, and CCR8 [13, 27, 36, 37]. The differential migration program of polarized T cells is to amplify certain types of immune responses (e.g. Th1 or Th2) most suitable to clear distinct types of pathogens (e.g. intracellular pathogens, extracellular bacteria or helminthes). The organ-specific migration is to promote regional immunity to limit the area of surveillance for each memory T cell and, therefore, allows the T cell to more effectively respond to pathogens.

Generation of the Functionally Specialized T-Helper Cell Subsets

Functionally specialized T-helper cells are the descendents of naive T cells as the result of antigen priming and T-cell differentiation. This occurs most desirably in SLT located throughout the body such as lymph nodes, spleen, tonsils and Peyer's patches. For activation of naive T cells, dendritic cells (DCs) fetch antigens from sites of infection to the T zone of SLT. Antigen peptides presented on the MHC molecules of DCs activate the T-cell receptors of naive T cells. Each antigen peptide/MHC complex generates its own specific signal in terms of affinity and activation strength, and therefore is a factor in lineage specification of T-helper cells. Other signals from DCs such as costimulatory molecules and cytokines play important roles in this activation process and ultimately determine the fate of the differentiated T-helper cells [38]. The cytokines and costimulatory molecules expressed by DCs and tissue cells are regulated by pathogen-associated molecular pattern receptor ligands such as toll-like receptor ligands [39–41]. Thus, the pathogen-derived signals and their interpretation by the host immune system are the determining factors for the fate of antigen-primed T-helper cells.

During antigen priming of naive T cells in SLT, cytokines such as IL-21 and IL-6 promote the generation of CXCR5$^+$ T_{FH} cells in vivo [42–45]. One caveat is that it is difficult to reproduce the induction of T_{FH} cells by IL-21 and IL-6 in vitro, suggesting a possible involvement of additional factors and cells. For generation of Th17 cells, IL-6, IL-21, TGF-β_1, and IL-23 are important [46–48]. For FoxP3$^+$ T cells, TGF-β_1 and IL-2 are important [49, 50]. Along with the cytokines, costimulatory molecules such as CD28, ICOS, 4-1BB, and OX40 are implicated in generation and/or maintenance of certain lineages of T-helper cells [51–55]. In addition, other signals such as retinoic acid promotes the generation of FoxP3$^+$ T cells [50], while S1P suppresses the generation and function of FoxP3$^+$ T cells [11, 56, 57]. The roles of cytokines in induction and effector function of the T-helper cells are depicted in figure 1. Many cytokines produced by a lineage of effector T cells can antagonize the differentiation of T-helper cells to other lineages. For example, IL-6, IL-4, and IFN-γ suppress the generation of FoxP3$^+$ T cells. IL-2, IL-4, and IFN-γ suppress the generation of Th17 cells. It is increasingly becoming clear that differentiation into a T-helper cell lineage is not the terminal destiny for the T cells. Th17 cells can become FoxP3$^+$ T cells and Th1 cells [58], FoxP3$^+$ T cells can become Th17 cells and T_{FH} cells depending on the cytokine milieu [59]. Early T_{FH} cells appear to be flexible in their differentiation, while the functionally mature GC T_{FH} cells are prone to die rather than to differentiate into other T-helper cells [60].

Migration and Function of T_{FH} Cells

T_{FH} cells are the T-helper cells found around or within the B-cell follicles and regulate the activation and maturation of B cells into effector or memory cells. Effector B cells are plasma cells that can produce specific types of antibodies (e.g. IgG1, IgG2, IgG3, IgG4, IgE and IgA). Memory B cells can quickly become plasma cells upon re-exposure to the same antigens. T_{FH} cells are defined by the expression of CXCR5. Its only ligand CXCL13 is specifically expressed in B-cell follicles [24, 61, 62]. T_{FH} cells appear heterogeneous in terms of tissue tropism, differentiation, and function. CXCR5$^+$CCR7$^+$ T_{FH} cells are early T_{FH} cells and reside in the T-B cell border area for T-B cell interaction at early stages of humoral immune response. Circulating CXCR5$^+$ T cells belong to this group. At a later stage, CXCR5$^+$CCR7$^-$ T_{FH} cells appear in germinal centers in secondary follicles. Thus, T_{FH} cells emerge first in the T-B border area and migrate into germinal centers as they differentiate. The T cells in the T-B border area migrate into germinal centers as a stable conjugate with B cells for persistent monogamous bidirectional activation [63]. SLAM-associated protein (SAP), an adaptor protein involved in the signaling of lymphocytic activation molecule (SLAM)

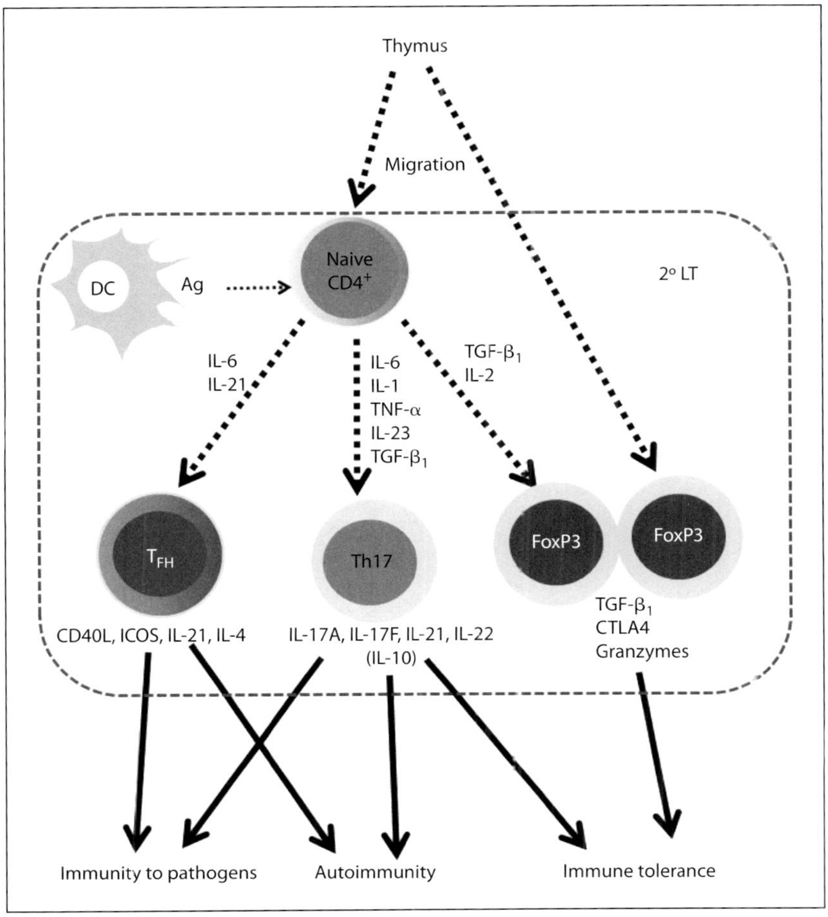

Fig. 1. Origin and function of the three T-helper cell subsets. All of the three cell subsets originate from naive T cells following antigen priming in SLT. Additionally, FoxP3+ T cells can be generated in the thymus and migrate into SLT. Naive T cells become Th17 cells when the tissue environment is rich with IL-6, TGF-β_1, IL-1, and IL-23 but is deficient with IFN-γ and IL-4. Naive ¯ cells become CXCR5+ T_{FH} cells in the presence of IL-21 and/or IL-6. FoxP3+ T cells are induced when there are TGF-β_1 and IL-2 but there is little of the cytokines that induce Th1, Th2, Th17 cells and T_{FH} cells. Th17 cells produce IL-17A, IL-17F, IL-21 and IL-22 and induce immune responses to fight against bacteria and fungi. If they are self-reactive, Th17 cells can induce severe tissue inflammation. T_{FH} cells provide the T-helper signals (CD40L, ICOS, IL-21 and IL-4) for activation of B cells at the T-B border and germinal centers. FoxP3+ T cells suppress antigen-presenting cells and effector T cells to protect the host from uncontrolled immune responses.

family of receptors [64], is required for the monogamous interaction between T and B cells [65]. Migration of T_{FH} cells is determined by the balance between CCR7 ligands (CCL19 and CCL21) and CXCR5 ligand (CXCL13) [66, 67]. CXCR5+CCR7+ T_{FH} cells migrate to the border and the mantle zone of the follicles because they express both CXCR5 and CCR7 and respond to both the T-cell zone and B-cell zone

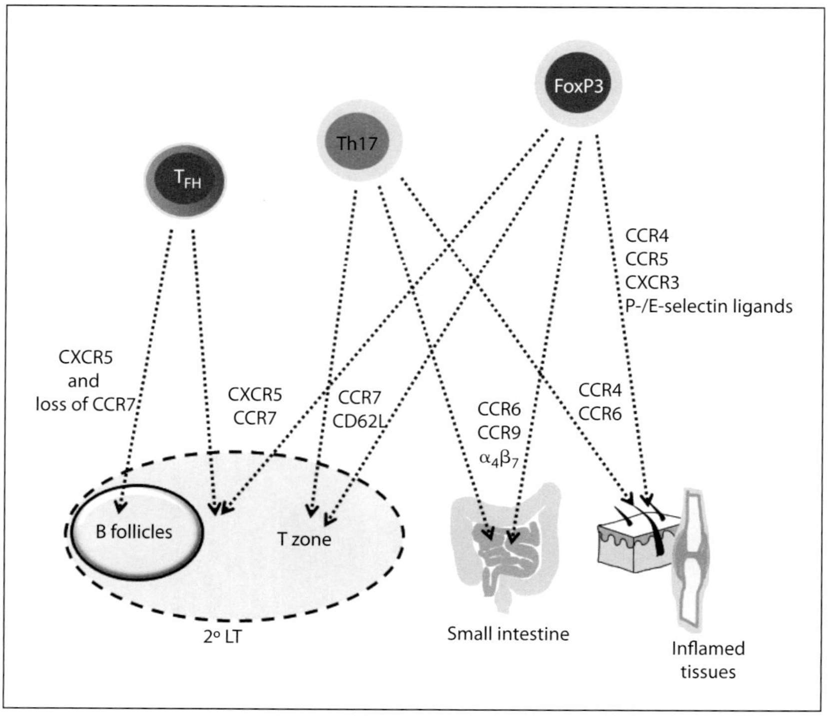

Fig. 2. Trafficking receptors of T_{FH} cells, Th17 cells, and FoxP3$^+$ T cells. Naive CD4+ T cells and FoxP3$^+$ T cells that are made in the thymus migrate to SLT using CD62L and CCR7. Naive CD4+ T cells become T_{FH} cells, Th17 cells, and FoxP3$^+$ T cells upon antigen priming, and memory/effector-related trafficking receptors are upregulated on the differentiated T-helper cells. CXCR5 guides T_{FH} cells into the T-B border area but not to germinal centers if they continue to express CCR7. Only fully differentiated T_{FH} cells lose CCR7 and maintain CXCR5 to stay within germinal centers. FoxP3$^+$ T cells can gain the expression of CXCR5 and migrate to the T-B border area to dampen the humoral immune responses. CXCR5$^+$ Th17 cells have the potential to migrate to the T-B border and positively affect the B-cell differentiation. CCR6 is the most characteristic receptor for Th17 cells. CCR6 appears to promote the migration of Th17 cells to the small intestine, peritoneal cavity, and inflamed tissues such as joints and central nervous tissues. Induction of CCR9 and $\alpha_4\beta_7$ is regulated by retinoic acid and can guide the T-helper cells to the small intestine. Additional receptors can be expressed on the T-helper cells and further modify the migration of the cells. FoxP3$^+$ T cells highly express most memory/effector-related trafficking receptors and can effectively migrate to any tissue sites where effector T cells migrate to.

chemoattractants. CXCR5$^+$CCR7$^-$ T_{FH} cells can migrate into germinal centers because they respond to CXCL13 but not to CCR7 ligands [24]. The trafficking receptors important for T_{FH} cells versus other effector T cells are summarized in figure 2.

Production of high-affinity antibodies is the result of B-cell proliferation, somatic mutation, and selection in germinal centers. It is well established that activation of the B-cell antigen receptors and selection by follicular DCs play critical roles in this process [68]. Recent research by many groups suggests that T_{FH} cells are required for robust production of isotype class-switched antibodies and autoimmunity [69,

70]. T_{FH} cells provide costimulation signals such as CD40L, ICOS, and OX40 [69–72] and cytokine signals such as IL-21, IL-4, and IFN-γ [73–76]. It has been established that these molecules are required for production of class-switched high-affinity antibodies. CD40L and CD40 provide indispensible signals to promote the humoral immunity as exemplified in X-linked hyper-IgM syndrome [77]. IL-21 is required for optimal production of antibodies by B cells [78]. ICOS is required for the germinal center response and T-cell-dependent B-cell response [79, 80]. IL-4 and IFN-γ are respectively implicated in the immunoglobulin class switch into Th2 (IgG4 and IgE) and Th1 (IgG1) type antibodies [81, 82]. The recent work on T_{FH} cells by several groups established that one important source of the B-cell stimulating molecules is the T_{FH} cells.

There are many questions regarding T_{FH} cells. It has been determined that human T_{FH} cells are mostly non-polarized T cells [23]. However, mouse T_{FH} cells contain many Th2 cells and some Th1 cells [73–76]. This could be due to a species difference or may be an experimental difference resulting from examining naturally arising T_{FH} cells in human tonsils versus experimentally induced T_{FH} cells in mouse lymphoid tissues. Also, there could be differences between SLT in terms of the T_{FH} phenotype (e.g. tonsils vs. peripheral lymph nodes). T_{FH} are apparently heterogeneous in surface phenotype and expression of CCR7 [24, 61, 62]. Are T_{FH} cells heterogeneous in function as well? It is likely that there would be a division of labor in helping B cells. Also, different T_{FH} cells would be generated in response to different pathogens or immune responses. Are T_{FH} cells stable in their lineage commitment or can easily become polarized effector T cells? Many $CXCR5^-$ T cells may not be the committed B-cell-helping T cells. They could be a population appearing transiently during immune responses as suggested [83–85]. However, GC T_{FH} cells seem to be fully committed T_{FH} cells that cannot become other effector T cells [60]. What is the transcription factor important for the lineage commitment for T_{FH} cells? BCL6 is highly expressed by human and mouse T_{FH} cells but the functional role of this transcription factor is unknown [86, 87]. STAT3 may be a factor important for relaying the signals of IL-21 and IL-6 in generation of B-cell-helping T cells but it does not induce the expression of CXCR5 [45].

Migration and Function of Th17 Cells

While T_{FH} cells stay and function within the SLT, most Th17 cells migrate out of the SLT in order to migrate into inflamed tissue sites. The frequency of Th17 cells in lymph nodes in the absence of apparent immune response is low (<0.2% of CD4+ T cells) [26]. The intestine is perhaps the most Th17 cell-enriched organ in the body without apparent inflammation (~5% of CD4+ T cells in the small intestine and ~2% in Peyer's patches) in the Balb/C mice housed at Purdue University. It has been reported that the frequencies of Th17 cells even in the same strain can vary in mice housed in different facilities [88]. This implies that the gut environment, formed by the gut microbiota, is an important factor for generation of Th17 cells in the intestine. In inflamed CNS and joints, Th17 cells become a major effector T population among CD4+ T cells [89, 90]. While it is assumed widely, there is no direct evidence that most of the Th17 cells enriched in inflamed tissues and intestine are the result of cell migration from the SLT. It is likely that some Th17 cells are induced in situ from pre-Th17 cells in inflamed tissues.

Th17 cells express a number of different trafficking receptors that are expressed by Th1 (e.g. CXCR3 and CCR5), Th2 (e.g. CCR4), T_{FH} (e.g. CXCR5) or $FoxP3^+$ (e.g. CCR4 and CCR6) T cells [25, 26]. Many Th17 cells express also lymphoid tissue homing receptors (CCR7 and CD62L). Thus, Th17 cells are likely to use all of these receptors to migrate to various tissue sites of inflammation and immune responses. For example, expression of CXCR5 by Th17 cells in certain lymphoid tissues would guide Th17 cells to B-cell follicles [26]. The trafficking receptors would also promote the interaction of Th17 cells with antigen-presenting cells or downstream effector cells. While most of these receptors are variably expressed by the Th17 cells of different organs, CCR6 is expressed by almost all of the Th17 cells in any mouse or human organs examined [25, 26, 48, 91, 92]. Thus, CCR6 is the receptor that best represents the Th17 cell lineage. TGF-β_1 is the cytokine that induces CCR6 in Th17 cells, while IL-2 is a cytokine that suppresses the expression of CCR6 in Th17 cells [26]. CCR6 is a homing receptor required for migration of Th17 cells to certain tissue sites in the intestine. Peyer's patches and other GALT are major expression sites of CCL20 [93–99], which is the ligand of CCR6. In this regard, Th17 cells migrate to the small intestine and peritoneal cavity in a CCR6-dependent manner [26]. CCR6 is also implicated in Th17 cell migration to inflamed joints and tissue sites of experimental autoimmune encephalomyelitis [92, 100]. An interesting feature of Th17 cells is that they express CCL20 at levels high enough to desensitize CCR6 [92, 100]. One can wonder if the CCL20 produced by Th17 cells can desensitize the CCR6 expressed by Th17 cells in an autocrine manner. This may further regulate the Th17 cell migration in response to additional competing trafficking signals.

Once within effector sites, Th17 cells can produce cytokines such as IL-17A, IL-17F, IL-21 and IL-22. IL-17A and IL-17F can activate tissue cells for expression of chemokines such as MCP-1, IP-10, IL-8 and cytokines such as IL-1β, TNF-α, IL-6, and G-CSF that recruit and activate neutrophils and other leukocytes [101–104]. IL-21 plays a role in maintenance of Th17 cells in an autocrine manner [105, 106]. As described earlier, IL-21 can boost humoral immune responses as well. However, the role of Th17 cell-derived IL-21 in regulation of humoral immune responses has not been determined yet. IL-22 can induce various antimicrobial proteins and a lipopolysaccharide-binding protein [107–110]. These effector cytokines can elicit not only antipathogenic responses but also inflammatory responses leading to tissue damage. For this reason, Th17 cell responses should be terminated later in the immune responses to prevent chronic inflammation. In this regard, many Th1 and

Th2 cytokines, produced by other effector T cells that are induced late in the immune responses, can suppress the responses mediated by Th17 cells [111]. It is plausible that Th17 cells transdifferentiate into other effector T cells once the need for Th17 cells subsides and the cytokine milieu is no longer conducive for maintenance of Th17 cells. Thus, in addition to migration, maintenance of existing Th17 cells would play an important role in determining the frequency of Th17 cells in a given tissue site.

Migration and Function of FoxP3$^+$ Cells

FoxP3$^+$ T cells are functionally distinct from T_{FH} cells and Th17 cells. While, T_{FH} cells and Th17 cells are effector cells important for immunity, FoxP3$^+$ T cells set a limit for differentiation and activation of conventional T cells including T_{FH} cells and Th17 cells. Therefore, they play important roles in protecting the host from potentially excessive immunological assaults. Our immune system is designed to produce FoxP3$^+$ T cells in both thymus and in the periphery. Thymically generated FoxP3$^+$ T cells (tFoxP3$^+$ T cells) exclusively migrate to the SLT and function to limit the activation of naive T cells and their differentiation into effector T cells [20, 112]. The migration of tFoxP3$^+$ T cells is thought to be mediated by CD62L and CCR7 in a manner similar to the migration of conventional T cells [113]. The fate of the naive T cell-like tFoxP3$^+$ T cells, entered into SLT, is determined by antigen signals and cytokine milieu. They become memory cell-like FoxP3$^+$ T cells following antigen priming and gain the expression of many non-LT homing receptors that are commonly expressed by Th1 (CCR5 and CXCR3), Th2 (CCR4), Th17 (CCR6), and T_{FH} (CXCR5) cells [20]. It appears that there is no trafficking receptor specifically expressed by the FoxP3$^+$ T-cell lineage. Rather, FoxP3$^+$ T cells express the trafficking receptors that are commonly expressed by conventional T cells. This is perhaps to codistribute the FoxP3$^+$ T cells together with other effector T cells for effective restraining of the activity of the effector T cells in most tissues.

An interesting feature of FoxP3$^+$ T cells is their accelerated acquisition of memory/effector-type trafficking receptors following antigen priming [20, 114]. In other words, FoxP3$^+$ T cells gain the expression of CCR4, CCR5, CCR6, CCR8, CXCR3, and CXCR6 more readily than do conventional T cells during antigen priming. Consistently, the migration of FoxP3$^+$ T cells to peripheral tissues is more efficient compared to conventional T cells [20]. Among the receptors expressed by FoxP3$^+$ T cells, CCR4 is required for suppression of inflammation in scurfy mice and of colitis in SCID mice by FoxP3$^+$ T cells [115, 116]. CCR7 and CD62L are implicated in prevention of graft-versus-host disease and colitis by FoxP3$^+$ T cells [117–119]. CCR7 is required for suppression of antigen-induced T-cell responses in lymph nodes [117]. CCR6 is implicated in migration of FoxP3$^+$ T cells to inflamed CNS [120]. CCR5 is important for their migration to Crohn's disease like lesions in the small intestine and in suppression of an experimental graft-versus-host disease in mice [121, 122].

The mechanisms that $FoxP3^+$ T cells would utilize to suppress target cells are still not clear. Generally, the suppression by $FoxP3^+$ T cells is dependent on cell-cell contact at least in vitro. It appears that TGF-β_1 is, perhaps, the most important effector cytokine that $FoxP3^+$ T cells use for their effector function [123]. $FoxP3^+$ T cells can kill target cells using granzymes [124–126]. $FoxP3^+$ T cells can suppress antigen-presenting cells employing the CTLA4 and indole-amine 2,3-dioxygenase pathway [127, 128]. In addition, carbon monoxide is implicated in the suppression mediated by $FoxP3^+$ T cells [129, 130]. Two-photon confocal microscopy revealed that $FoxP3^+$ T cells suppress the interaction of conventional T cells with DCs during antigen priming in SLT [131]. Thus, $FoxP3^+$ T cells appear to inhibit the establishment of stable contacts between naive T cells by DCs. Also, in suppression of cytotoxic T cells, $FoxP3^+$ T cells do not seem to fully suppress the differentiation process of CD8 T cells in vivo [132]. Rather, $FoxP3^+$ T cells conditionally suppress the exocytosis of cytotoxic granules without directly contacting with the target T cells. It appears that $FoxP3^+$ T cells interact preferentially with DCs than with conventional T cells to exert their suppressive activity [133]. Thus, in vivo, the trafficking signals that colocalize $FoxP3^+$ T cells and antigen-presenting cells would be important for indirectly suppressing target T cells.

$FoxP3^+$ T cells can play critical roles in regulation of the humoral immune response by migrating to the T-B cell border and germinal centers to dampen the B-cell differentiation process [18]. CXCR5 and its ligand CXCL13 are implicated in this process. There are considerable numbers of $CXCR5^+$ $FoxP3^+$ T cells in the area surrounding and within the mantle zone. Some $FoxP3^+$ T cells are found within germinal centers. $FoxP3^+$ T cells can directly suppress both T_{FH} cells and B cells [134]. It is generally thought that $FoxP3^+$ T cells are relatively stable in their lineage commitment. However, a recent report suggests that $FoxP3^+$ T cells can become T_{FH} cells within Peyer's patches [135]. There have been reports that $FoxP3^+$ T cells can convert to Th17 cells and other effector T cells, suggesting that the $FoxP3^+$ T cells are not immune to conversion to other lineages [59].

Concluding Remarks

T_{FH} cells and Th17 cells are effector T cells but they are distinct in migratory behavior and effector function. T_{FH} cells localize in the follicles within SLT for regulation of B-cell maturation to plasma cells and memory B cells. Th17 cells migrate to the periphery for regulation of antimicrobial responses but can cause chronic inflammation if they persist in the tissues. In contrast, $FoxP3^+$ T cells play essential roles in reigning in generation and function of the effector T cells in both SLT and peripheral tissues. While the trafficking receptors expressed by the two effector T-cell subsets appear distinct, $FoxP3^+$ T cells are versatile in expression of trafficking receptors so that they can gain access to all of the tissues harboring effector T cells (fig. 2). Therefore, the migration and expression of trafficking receptors by the three T-cell subsets is designed to effectively achieve both immunity and immune tolerance.

References

1 Harrington LE, Mangan PR, Weaver CT: Expanding the effector CD4 T-cell repertoire: the Th17 lineage. Curr Opin Immunol 2006;18:349–356.

2 Schaerli P, Willimann K, Lang AB, Lipp M, Loetscher P, Moser B: CXC chemokine receptor 5 expression defines follicular homing T cells with B-cell helper function. J Exp Med 2000;192:1553–1562.

3 Breitfeld D, Ohl L, Kremmer E, Ellwart J, Sallusto F, Lipp M, Forster R: Follicular B helper T cells express CXC chemokine receptor 5, localize to B-cell follicles, and support immunoglobulin production. J Exp Med 2000;192:1545–1552.

4 Kim CH, Rott LS, Clark-Lewis I, Campbell DJ, Wu L, Butcher EC: Subspecialization of CXCR5+ T cells: B helper activity is focused in a germinal center-localized subset of CXCR5+ T cells. J Exp Med 2001;193:1373–1381.

5 Kim CH: Migration and function of FoxP3+ regulatory T cells in the hematolymphoid system. Exp Hematol 2006;34:1033–1040.

6 Park H, Li Z, Yang XO, Chang SH, Nurieva R, Wang YH, Wang Y, Hood L, Zhu Z, Tian Q, Dong C: A distinct lineage of CD4 T cells regulates tissue inflammation by producing interleukin-17. Nat Immunol 2005;6:1133–1141.

7 Yu JJ, Gaffen SL: Interleukin-17: a novel inflammatory cytokine that bridges innate and adaptive immunity. Front Biosci 2008;13:170–177.

8 McHeyzer-Williams LJ, Pelletier N, Mark L, Fazilleau N, McHeyzer-Williams MG: Follicular helper T cells as cognate regulators of B-cell immunity. Curr Opin Immunol 2009;21:266–273.

9 Shevach EM: Mechanisms of FoxP3+ T-regulatory cell-mediated suppression. Immunity 2009;30:636–645.

10 Goetzl EJ, Liao JJ, Huang MC: Regulation of the roles of sphingosine 1-phosphate and its type 1 G protein-coupled receptor in T cell immunity and autoimmunity. Biochim Biophys Acta 2008;1781:503–507.

11 Schwab SR, Pereira JP, Matloubian M, Xu Y, Huang Y, Cyster JG: Lymphocyte sequestration through S1P lyase inhibition and disruption of S1P gradients. Science 2005;309:1735–1739.

12 Schwab SR, Cyster JG: Finding a way out: lymphocyte egress from lymphoid organs. Nat Immunol 2007;8:1295–1301.

13 Marelli-Berg FM, Cannella L, Dazzi F, Mirenda V: The highway code of T cell trafficking. J Pathol 2008;214:179–189.

14 Brinkmann V, Davis MD, Heise CE, Albert R, Cottens S, Hof R, Bruns C, Prieschl E, Baumruker T, Hiestand P, Foster CA, Zollinger M, Lynch KR: The immune modulator FTY720 targets sphingosine 1-phosphate receptors. J Biol Chem 2002;277:21453–21457.

15 Matloubian M, Lo CG, Cinamon G, Lesneski MJ, Xu Y, Brinkmann V, Allende ML, Proia RL, Cyster JG: Lymphocyte egress from thymus and peripheral lymphoid organs is dependent on S1P receptor 1. Nature 2004;427:355–360.

16 Allende ML, Dreier JL, Mandala S, Proia RL: Expression of the sphingosine 1-phosphate receptor, S1P1, on T-cells controls thymic emigration. J Biol Chem 2004;279:15396–15401.

17 Lim HW, Broxmeyer HE, Kim CH: Regulation of trafficking receptor expression in human forkhead box P3+ regulatory T cells. J Immunol 2006;177:840–851.

18 Lim HW, Hillsamer P, Kim CH: Regulatory T cells can migrate to follicles upon T cell activation and suppress GC-Th cells and GC-Th cell-driven B-cell responses. J Clin Invest 2004;114:1640–1649.

19 Butcher EC, Williams M, Youngman K, Rott L, Briskin M: Lymphocyte trafficking and regional immunity. Adv Immunol 1999;72:209–253.

20 Lee JH, Kang SG, Kim CH: FoxP3+ T cells undergo conventional first switch to lymphoid tissue homing receptors in thymus but accelerated second switch to non-lymphoid tissue homing receptors in secondary lymphoid tissues. J Immunol 2007;178:301–311.

21 Kim CH: The greater chemotactic network for lymphocyte trafficking: chemokines and beyond. Curr Opin Hematol 2005;12:298–304.

22 Austrup F, Vestweber D, Borges E, Lohning M, Brauer R, Herz U, Renz H, Hallmann R, Scheffold A, Radbruch A, Hamann A: P- and E-selectin mediate recruitment of T-helper-1 but not T-helper-2 cells into inflamed tissues. Nature 1997;385:81–83.

23 Kim CH, Campbell DJ, Butcher EC: Nonpolarized memory T cells. Trends Immunol 2001;22:527–530.

24 Kim CH, Rott LS, Clark-Lewis I, Campbell DJ, Wu L, Butcher EC: Subspecialization of CXCR5+ T cells: B helper activity is focused in a germinal center-localized subset of CXCR5+ T cells. J Exp Med 2001;193:1373–1381.

25 Lim HW, Lee J, Hillsamer P, Kim CH: Human Th17 cells share major trafficking receptors with both polarized effector T cells and FoxP3+ regulatory T cells. J Immunol 2008;180:122–129.

26 Wang C, Kang SG, Lee J, Sun Z, Kim CH: The roles of CCR6 in migration of Th17 cells and regulation of effector T-cell balance in the gut. Mucosal Immunol 2009;2:173–183.
27 Kim CH: Chemokine-chemokine receptor network in immune cell trafficking. Curr Drug Targets Immune Endocr Metabol Disord 2004;4:343–361.
28 Rott LS, Rose JR, Bass D, Williams MB, Greenberg HB, Butcher EC: Expression of mucosal homing receptor $\alpha_4\beta_7$ by circulating CD4+ cells with memory for intestinal rotavirus. J Clin Invest 1997;100: 1204–1208.
29 Williams MB, Butcher EC: Homing of naive and memory T lymphocyte subsets to Peyer's patches, lymph nodes, and spleen. J Immunol 1997;159:1746–1752.
30 Bradley LM, Malo ME, Fong S, Tonkonogy SL, Watson SR: Blockade of both L-selectin and α_4 integrins abrogates naive CD4 cell trafficking and responses in gut-associated lymphoid organs. Int Immunol 1998;10:961–968.
31 Stenstad H, Ericsson A, Johansson-Lindbom B, Svensson M, Marsal J, Mack M, Picarella D, Soler D, Marquez G, Briskin M, Agace WW: Gut-associated lymphoid tissue-primed CD4+ T cells display CCR9-dependent and -independent homing to the small intestine. Blood 2006;107:3447–3454.
32 Svensson M, Marsal J, Ericsson A, Carramolino L, Broden T, Marquez G, Agace WW: CCL25 mediates the localization of recently activated CD8αβ+ lymphocytes to the small-intestinal mucosa. J Clin Invest 2002;110:1113–1121.
33 Wurbel MA, Malissen M, Guy-Grand D, Meffre E, Nussenzweig MC, Richelme M, Carrier A, Malissen B: Mice lacking the CCR9 CC-chemokine receptor show a mild impairment of early T- and B-cell development and a reduction in T-cell receptor $\gamma\delta^+$ gut intraepithelial lymphocytes. Blood 2001;98: 2626–2632.
34 Papadakis KA, Prehn J, Nelson V, Cheng L, Binder SW, Ponath PD, Andrew DP, Targan SR: The role of thymus-expressed chemokine and its receptor CCR9 on lymphocytes in the regional specialization of the mucosal immune system. J Immunol 2000; 165:5069–5076.
35 Kunkel EJ, Campbell JJ, Haraldsen G, Pan J, Boisvert J, Roberts AI, Ebert EC, Vierra MA, Goodman SB, Genovese MC, Wardlaw AJ, Greenberg HB, Parker CM, Butcher EC, Andrew DP, Agace WW: Lymphocyte CC chemokine receptor 9 and epithelial thymus-expressed chemokine expression distinguish the small intestinal immune compartment: epithelial expression of tissue-specific chemokines as an organizing principle in regional immunity. J Exp Med 2000;192:761–768.
36 Hwang ST: Mechanisms of T-cell homing to skin. Adv Dermatol 2001;17:211–241.
37 Schaerli P, Ebert L, Willimann K, Blaser A, Roos RS, Loetscher P, Moser B: A skin-selective homing mechanism for human immune surveillance T cells. J Exp Med 2004;199:1265–1275.
38 Murphy KM: Fate vs. choice: the immune system reloaded. Immunol Res 2005;32:193–200.
39 Netea MG, Sutmuller R, Hermann C, Van der Graaf CA, Van der Meer JW, van Krieken JH, Hartung T, Adema G, Kullberg BJ: Toll-like receptor 2 suppresses immunity against *Candida albicans* through induction of IL-10 and regulatory T cells. J Immunol 2004;172:3712–3718.
40 Armant MA, Fenton MJ: Toll-like receptors: a family of pattern-recognition receptors in mammals. Genome Biol 2002;3:Reviews3011.
41 Akira S, Takeda K, Kaisho T: Toll-like receptors: critical proteins linking innate and acquired immunity. Nat Immunol 2001;2:675–680.
42 Bauquet AT, Jin H, Paterson AM, Mitsdoerffer M, Ho IC, Sharpe AH, Kuchroo VK: The costimulatory molecule ICOS regulates the expression of c-Maf and IL-21 in the development of follicular T-helper cells and TH-17 cells. Nat Immunol 2009;10:167–175.
43 Vogelzang A, McGuire HM, Yu D, Sprent J, Mackay CR, King C: A fundamental role for interleukin-21 in the generation of T follicular helper cells. Immunity 2008;29:127–137.
44 Nurieva RI, Chung Y, Hwang D, Yang XO, Kang HS, Ma L, Wang YH, Watowich SS, Jetten AM, Tian Q, Dong C: Generation of T follicular helper cells is mediated by interleukin-21 but independent of T-helper 1, 2, or 17 cell lineages. Immunity 2008;29: 138–149.
45 Eddahri F, Denanglaire S, Bureau F, Spolski R, Leonard WJ, Leo O, Andris F: Interleukin-6/STAT3 signaling regulates the ability of naive T cells to acquire B-cell help capacities. Blood 2009;113:2426–2433.
46 Wei L, Laurence A, Elias KM, O'Shea JJ: IL-21 is produced by Th17 cells and drives IL-17 production in a STAT3-dependent manner. J Biol Chem 2007;282:34605–34610.
47 Kimura A, Naka T, Kishimoto T: IL-6-dependent and -independent pathways in the development of interleukin 17-producing T-helper cells. Proc Natl Acad Sci USA 2007;104:12099–12104.
48 Annunziato F, Cosmi L, Santarlasci V, Maggi L, Liotta F, Mazzinghi B, Parente E, Fili L, Ferri S, Frosali F, Giudici F, Romagnani P, Parronchi P, Tonelli F, Maggi E, Romagnani S: Phenotypic and functional features of human Th17 cells. J Exp Med 2007;204:1849–1861.

49 Kim CH: Molecular targets of FoxP3⁺ regulatory T cells. Mini Rev Med Chem 2007;7:1136–1143.
50 Kim CH: Regulation of FoxP3 regulatory T cells and Th17 cells by retinoids. Clin Dev Immunol 2008; 2008:416910.
51 So T, Lee SW, Croft M: Immune regulation and control of regulatory T cells by OX40 and 4–1BB. Cytokine Growth Factor Rev 2008;19:253–262.
52 Van Berkel ME, Oosterwegel MA: CD28 and ICOS: similar or separate costimulators of T cells? Immunol Lett 2006;105:115–122.
53 Croft M: Costimulation of T cells by OX40, 4-1BB, and CD27. Cytokine Growth Factor Rev 2003;14: 265–273.
54 Wang C, Lin GH, McPherson AJ, Watts TH: Immune regulation by 4-1BB and 4-1BBL: complexities and challenges. Immunol Rev 2009;229: 192–215.
55 Lane P: Role of OX40 signals in coordinating CD4 T cell selection, migration, and cytokine differentiation in Th1 and Th2 cells. J Exp Med 2000;191:201–206.
56 Sawicka E, Dubois G, Jarai G, Edwards M, Thomas M, Nicholls A, Albert R, Newson C, Brinkmann V, Walker C: The sphingosine 1-phosphate receptor agonist FTY720 differentially affects the sequestration of CD4+/CD25+ T-regulatory cells and enhances their functional activity. J Immunol 2005; 175:7973–7980.
57 Liu G, Burns S, Huang G, Boyd K, Proia RL, Flavell RA, Chi H: The receptor S1P1 overrides regulatory T cell-mediated immune suppression through Akt-mTOR. Nat Immunol 2009;
58 Lee YK, Mukasa R, Hatton RD, Weaver CT: Developmental plasticity of Th17 and Treg cells. Curr Opin Immunol 2009;21:274–280.
59 Zhou X, Bailey-Bucktrout S, Jeker LT, Bluestone JA: Plasticity of CD4+ FoxP3⁺ T cells. Curr Opin Immunol 2009;21:281–285.
60 Lim HW, Kim CH: Loss of IL-7 receptor α on CD4+ T cells defines terminally differentiated B-cell-helping effector T cells in a B-cell-rich lymphoid tissue. J Immunol 2007;179:7448–7456.
61 Breitfeld D, Ohl L, Kremmer E, Ellwart J, Sallusto F, Lipp M, Forster R: Follicular B helper T cells express CXC chemokine receptor 5, localize to B-cell follicles, and support immunoglobulin production. J Exp Med 2000;192:1545–1552.
62 Schaerli P, Willimann K, Lang AB, Lipp M, Loetscher P, Moser B: CXC chemokine receptor 5 expression defines follicular homing T cells with B-cell helper function. J Exp Med 2000;192:1553–1562.
63 Okada T, Miller MJ, Parker I, Krummel MF, Neighbors M, Hartley SB, O'Garra A, Cahalan MD, Cyster JG: Antigen-engaged B-cells undergo chemotaxis toward the T zone and form motile conjugates with helper T cells. PLoS Biol 2005;3:e150.
64 Latour S, Veillette A: The SAP family of adaptors in immune regulation. Semin Immunol 2004;16:409–419.
65 Qi H, Cannons JL, Klauschen F, Schwartzberg PL, Germain RN: SAP-controlled T-B-cell interactions underlie germinal centre formation. Nature 2008;455:764–769.
66 Haynes NM, Allen CD, Lesley R, Ansel KM, Killeen N, Cyster JG: Role of CXCR5 and CCR7 in follicular Th cell positioning and appearance of a programmed cell death gene-1(high) germinal center-associated subpopulation. J Immunol 2007;179:5099–5108.
67 Hardtke S, Ohl L, Forster R: Balanced expression of CXCR5 and CCR7 on follicular T-helper cells determines their transient positioning to lymph node follicles and is essential for efficient B-cell help. Blood 2005;106:1924–1931.
68 Cyster JG, Ansel KM, Reif K, Ekland EH, Hyman PL, Tang HL, Luther SA Ngo VN: Follicular stromal cells and lymphocyte homing to follicles. Immunol Rev 2000;176:181–193.
69 Kim JR, Lim HW, Kang SG, Hillsamer P, Kim CH: Human CD57+ germinal center-T cells are the major helpers for GC-B cells and induce class switch recombination. BMC Immunol 2005;6:3.
70 Odegard JM, Marks BR DiPlacido LD, Poholek AC, Kono DH, Dong C, Flavell RA, Craft J: ICOS-dependent extrafollicular helper T cells elicit IgG production via IL-21 in systemic autoimmunity. J Exp Med 2008;205:2873–2886.
71 Obermeier F, Schwarz H, Dunger N, Strauch UG, Grunwald N, Scholmerich J, Falk W: OX40/OX40L interaction induces the expression of CXCR5 and contributes to chronic colitis induced by dextran sulfate sodium in mice. Eur J Immunol 2003;33: 3265–3274.
72 Walker LS, Gulbranson-Judge A, Flynn S, Brocker T, Raykundalia C, Goodall M, Forster R, Lipp M, Lane P: Compromised OX40 function in CD28-deficient mice is linked with failure to develop CXC chemokine receptor 5-positive CD4 cells and germinal centers. J Exp Med 1999;190:1115–1122.
73 Bryant VL, Ma CS, Avery DT, Li Y, Good KL, Corcoran LM, de Waal Malefyt R, Tangye SG: Cytokine-mediated regulation of human B-cell differentiation into Ig-secreting cells: predominant role of IL-21 produced by CXCR5⁺ T follicular helper cells. J Immunol 2007;179:8180–8190.
74 Reinhardt RL, Liang HE, Locksley RM: Cytokine-secreting follicular T cells shape the antibody repertoire. Nat Immunol 2009;10:385–393.

75 Zaretsky AG, Taylor JJ, King IL, Marshall FA, Mohrs M, Pearce EJ: T follicular helper cells differentiate from Th2 cells in response to helminth antigens. J Exp Med 2009;206:991–999.
76 King IL, Mohrs M: IL-4-producing CD4+ T cells in reactive lymph nodes during helminth infection are T follicular helper cells. J Exp Med 2009;206:1001–1007.
77 Aruffo A, Farrington M, Hollenbaugh D, Li X, Milatovich A, Nonoyama S, Bajorath J, Grosmaire LS, Stenkamp R, Neubauer M, et al: The CD40 ligand, gp39, is defective in activated T cells from patients with X-linked hyper-IgM syndrome. Cell 1993;72:291–300.
78 Ozaki K, Spolski R, Feng CG, Qi CF, Cheng J, Sher A, Morse HC 3rd, Liu C, Schwartzberg PL, Leonard WJ: A critical role for IL-21 in regulating immunoglobulin production. Science 2002;298:1630–1634.
79 Tafuri A, Shahinian A, Bladt F, Yoshinaga SK, Jordana M, Wakeham A, Boucher LM, Bouchard D, Chan VS, Duncan G, Odermatt B, Ho A, Itie A, Horan T, Whoriskey JS, Pawson T, Penninger JM, Ohashi PS, Mak TW: ICOS is essential for effective T-helper-cell responses. Nature 2001;409:105–109.
80 Dong C, Temann UA, Flavell RA: Cutting edge: critical role of inducible costimulator in germinal center reactions. J Immunol 2001;166:3659–3662.
81 Elenich LA, Ford CS, Collins JT, Dunnick WA: Gamma-1 heavy chain transgenes are responsive to IFN-γ repression and CD40 ligation. J Immunol 1997;158:4564–4573.
82 Purkerson J, Isakson P: A two-signal model for regulation of immunoglobulin isotype switching. FASEB J 1992;6:3245–3252.
83 Schaerli P, Loetscher P, Moser B: Cutting edge: induction of follicular homing precedes effector Th cell development. J Immunol 2001;167:6082–6086.
84 Ansel KM, McHeyzer-Williams LJ, Ngo VN, McHeyzer-Williams MG, Cyster JG: In vivo-activated CD4 T cells upregulate CXC chemokine receptor 5 and reprogram their response to lymphoid chemokines. J Exp Med 1999;190:1123–1134.
85 Kim CH, Nagata K, Butcher EC: Dendritic cells support sequential reprogramming of chemoattractant receptor profiles during naive to effector T cell differentiation. J Immunol 2003;171:152–158.
86 Chtanova T, Tangye SG, Newton R, Frank N, Hodge MR, Rolph MS, Mackay CR: T follicular helper cells express a distinctive transcriptional profile, reflecting their role as non-Th1/Th2 effector cells that provide help for B cells. J Immunol 2004;173:68–78.
87 Kim CH, Lim HW, Kim JR, Rott L, Hillsamer P, Butcher EC: Unique gene expression program of human germinal center T-helper cells. Blood 2004;104:1952–1960.
88 Ivanov II, Frutos Rde L, Manel N, Yoshinaga K, Rifkin DB, Sartor RB, Finlay BB, Littman DR: Specific microbiota direct the differentiation of IL-17-producing T-helper cells in the mucosa of the small intestine. Cell Host Microbe 2008;4:337–349.
89 Sato K, Suematsu A, Okamoto K, Yamaguchi A, Morishita Y, Kadono Y, Tanaka S, Kodama T, Akira S, Iwakura Y, Cua DJ, Takayanagi H: Th17 functions as an osteoclastogenic helper T cell subset that links T cell activation and bone destruction. J Exp Med 2006;203:2673–2682.
90 Sutton C, Brereton C, Keogh B, Mills KH, Lavelle EC: A crucial role for IL-1 in the induction of IL-17-producing T cells that mediate autoimmune encephalomyelitis. J Exp Med 2006;203:1685–1691.
91 Singh SP, Zhang HH, Foley JF, Hedrick MN, Farber JM: Human T cells that are able to produce IL-17 express the chemokine receptor CCR6. J Immunol 2008;180:214–221.
92 Hirota K, Yoshitomi H, Hashimoto M, Maeda S, Teradaira S, Sugimoto N, Yamaguchi T, Nomura T, Ito H, Nakamura T, Sakaguchi N, Sakaguchi S: Preferential recruitment of CCR6-expressing Th17 cells to inflamed joints via CCL20 in rheumatoid arthritis and its animal model. J Exp Med 2007;204:2803–2812.
93 Kwon JH, Keates S, Bassani L, Mayer LF, Keates AC: Colonic epithelial cells are a major site of macrophage inflammatory protein 3α production in normal colon and inflammatory bowel disease. Gut 2002;51:818–826.
94 Sierro F, Dubois B, Coste A, Kaiserlian D, Kraehenbuhl JP, Sirard JC: Flagellin stimulation of intestinal epithelial cells triggers CCL20-mediated migration of dendritic cells. Proc Natl Acad Sci USA 2001;98:13722–13727.
95 Fujiie S, Hieshima K, Izawa D, Nakayama T, Fujisawa R, Ohyanagi H, Yoshie O: Proinflammatory cytokines induce liver and activation-regulated chemokine/macrophage inflammatory protein-3α/CCL20 in mucosal epithelial cells through NF-κB. Int Immunol 2001;13:1255–1263.
96 Izadpanah A, Dwinell MB, Eckmann L, Varki NM, Kagnoff MF: Regulated MIP-3α/CCL20 production by human intestinal epithelium: mechanism for modulating mucosal immunity. Am J Physiol 2001;280:G710–719.
97 Nakayama T, Fujisawa R, Yamada H, Horikawa T, Kawasaki H, Hieshima K, Izawa D, Fujiie S, Tezuka T, Yoshie O: Inducible expression of a CC chemokine liver- and activation-regulated chemokine/macrophage inflammatory protein-3 α/CCL20 by epidermal keratinocytes and its role in atopic dermatitis. Int Immunol 2001;13:95–103.

98 Tanaka Y, Imai T, Baba M, Ishikawa I, Uehira M, Nomiyama H, Yoshie O: Selective expression of liver and activation-regulated chemokine in intestinal epithelium in mice and humans. Eur J Immunol 1999;29:633–642.

99 Varona R, Zaballos A, Gutierrez J, Martin P, Roncal F, Albar JP, Ardavin C, Marquez G: Molecular cloning, functional characterization and mRNA expression analysis of the murine chemokine receptor CCR6 and its specific ligand MIP-3α. FEBS Lett 1998;440:188–194.

100 Yamazaki T, Yang XO, Chung Y, Fukunaga A, Nurieva R, Pappu B, Martin-Orozco N, Kang HS, Ma L, Panopoulos AD, Craig S, Watowich SS, Jetten AM, Tian Q, Dong C: CCR6 regulates the migration of inflammatory and regulatory T cells. J Immunol 2008;181:8391–8401.

101 Kolls JK, Linden A: Interleukin-17 family members and inflammation. Immunity 2004;21:467–476.

102 Fossiez F, Djossou O, Chomarat P, Flores-Romo L, Ait-Yahia S, Maat C, Pin JJ, Garrone P, Garcia E, Saeland S, Blanchard D, Gaillard C, Das Mahapatra B, Rouvier E, Golstein P, Banchereau J, Lebecque S: T cell interleukin-17 induces stromal cells to produce proinflammatory and hematopoietic cytokines. J Exp Med 1996;183:2593–2603.

103 Kawaguchi M, Kokubu F, Odaka M, Watanabe S, Suzuki S, Ieki K, Matsukura S, Kurokawa M, Adachi M, Huang SK: Induction of granulocyte-macrophage colony-stimulating factor by a new cytokine, ML-1 (IL-17F), via Raf I-MEK-ERK pathway. J Allergy Clin Immunol 2004;114:444–450.

104 Henness S, Johnson CK, Ge Q, Armour CL, Hughes JM, Ammit AJ: IL-17A augments TNF-α-induced IL-6 expression in airway smooth muscle by enhancing mRNA stability. J Allergy Clin Immunol 2004; 114:958–964.

105 Korn T, Bettelli E, Gao W, Awasthi A, Jager A, Strom TB, Oukka M, Kuchroo VK: IL-21 initiates an alternative pathway to induce proinflammatory T_H17 cells. Nature 2007;448:484–487.

106 Nurieva R, Yang XO, Martinez G, Zhang Y, Panopoulos AD, Ma L, Schluns K, Tian Q, Watowich SS, Jetten AM, Dong C: Essential autocrine regulation by IL-21 in the generation of inflammatory T cells. Nature 2007;448:480–483.

107 Wolk K, Witte E, Wallace E, Docke WD, Kunz S, Asadullah K, Volk HD, Sterry W, Sabat R: IL-22 regulates the expression of genes responsible for antimicrobial defense, cellular differentiation, and mobility in keratinocytes: a potential role in psoriasis. Eur J Immunol 2006;36:1309–1323.

108 Nagalakshmi ML, Rascle A, Zurawski S, Menon S, de Waal Malefyt R: Interleukin-22 activates STAT3 and induces IL-10 by colon epithelial cells. Int Immunopharmacol 2004;4:679–691.

109 Radaeva S, Sun R, Pan HN, Hong F, Gao B: IL-22 plays a protective role in T cell-mediated murine hepatitis: IL-22 is a survival factor for hepatocytes via STAT3 activation. Hepatology 2004;39:1332–1342.

110 Wolk K, Witte E, Hoffmann U, Doecke WD, Endesfelder S, Asadullah K, Sterry W, Volk HD, Wittig BM, Sabat R: IL-22 induces lipopolysaccharide-binding protein in hepatocytes: a potential systemic role of IL-22 in Crohn's disease. J Immunol 2007;178:5973–5981.

111 Zhou L, Chong MM, Littman DR: Plasticity of CD4+ T cell lineage differentiation. Immunity 2009; 30:646–655.

112 Xu D, Liu H, Komai-Koma M, Campbell C, McSharry C, Alexander J, Liew FY: CD4+CD25+ regulatory T cells suppress differentiation and functions of Th1 and Th2 cells, *Leishmania major* infection, and colitis in mice. J Immunol 2003;170: 394–399.

113 Menning A, Hopken UE, Siegmund K, Lipp M, Hamann A, Huehn J: Distinctive role of CCR7 in migration and functional activity of naive- and effector/memory-like Treg subsets. Eur J Immunol 2007;37:1575–1583.

114 Lim HW, Broxmeyer HE, Kim CH: Regulation of trafficking receptor expression in human FOXP3+ regulatory T cells. J Immunol 2006;177:840–851.

115 Lee I, Wang L, Wells AD, Dorf ME, Ozkaynak E, Hancock WW: Recruitment of FoxP3+ T-regulatory cells mediating allograft tolerance depends on the CCR4 chemokine receptor. J Exp Med 2005;201: 1037–1044.

116 Sather BD, Treuting P, Perdue N, Miazgowicz M, Fontenot JD, Rudensky AY, Campbell DJ: Altering the distribution of Foxp3+ regulatory T cells results in tissue-specific inflammatory disease. J Exp Med 2007;204:1335–1347.

117 Schneider MA, Meingassner JG, Lipp M, Moore HD, Rot A: CCR7 is required for the in vivo function of CD4+ CD25− regulatory T cells. J Exp Med 2007;204:735–745.

118 Ermann J, Hoffmann P, Edinger M, Dutt S, Blankenberg FG, Higgins JP, Negrin RS, Fathman CG, Strober S: Only the CD62L+ subpopulation of CD4+CD25+ regulatory T cells protects from lethal acute GVHD. Blood 2005;105:2220–2226.

119 Taylor PA, Panoskaltsis-Mortari A, Swedin JM, Lucas PJ, Gress RE, Levine BL, June CH, Serody JS, Blazar BR: L-selectin(hi) but not the L-selectin(lo) CD4+25+ T-regulatory cells are potent inhibitors of GvHD and BM graft rejection. Blood 2004;104:3804–3812.

120 Villares R, Cadenas V, Lozano M, Almonacid L, Zaballos A, Martinez AC, Varona R: CCR6 regulates EAE pathogenesis by controlling regulatory CD4+ T-cell recruitment to target tissues. Eur J Immunol 2009;39:1671–1681.

121 Kang SG, Piniecki RJ, Hogenesch H, Lim HW, Wiebke E, Braun SE, Matsumoto S, Kim CH: Identification of a chemokine network that recruits FoxP3+ regulatory T cells into chronically inflamed intestine. Gastroenterology 2007;132:966–981.
122 Wysocki CA, Jiang Q, Panoskaltsis-Mortari A, Taylor PA, McKinnon KP, Su L, Blazar BR, Serody JS: Critical role for CCR5 in the function of donor CD4+CD25+ regulatory T cells during acute graft-versus-host disease. Blood 2005;106:3300–3307.
123 Nakamura K, Kitani A, Strober W: Cell contact-dependent immunosuppression by CD4+CD25+ regulatory T cells is mediated by cell surface-bound transforming growth factor β. J Exp Med 2001;194: 629–644.
124 Grossman WJ, Verbsky JW, Barchet W, Colonna M, Atkinson JP, Ley TJ: Human T-regulatory cells can use the perforin pathway to cause autologous target cell death. Immunity 2004;21:589–601.
125 Gondek DC, Lu LF, Quezada SA, Sakaguchi S, Noelle RJ: Cutting edge: contact-mediated suppression by CD4+CD25+ regulatory cells involves a granzyme B-dependent, perforin-independent mechanism. J Immunol 2005;174:1783–1786.
126 Zhao DM, Thornton AM, DiPaolo RJ, Shevach EM: Activated CD4+CD25+ T cells selectively kill B lymphocytes. Blood 2006;107:3925–3932.
127 Annunziato F, Cosmi L, Liotta F, Lazzeri E, Manetti R, Vanini V, Romagnani P, Maggi E, Romagnani S: Phenotype, localization, and mechanism of suppression of CD4+CD25+ human thymocytes. J Exp Med 2002;196:379–387.
128 Manzotti CN, Tipping H, Perry LC, Mead KI, Blair PJ, Zheng Y, Sansom DM: Inhibition of human T cell proliferation by CTLA-4 utilizes CD80 and requires CD25+ regulatory T cells. Eur J Immunol 2002;32:2888–2896.
129 Choi BM, Pae HO, Jeong YR, Kim YM, Chung HT: Critical role of heme oxygenase-1 in Foxp3-mediated immune suppression. Biochem Biophys Res Commun 2005;327:1066–1071.
130 Otterbein LE, Soares MP, Yamashita K, Bach FH: Heme oxygenase-1: unleashing the protective properties of heme. Trends Immunol 2003;24:449–455.
131 Tadokoro CE, Shakhar G, Shen S, Ding Y, Lino AC, Maraver A, Lafaille JJ, Dustin ML: Regulatory T cells inhibit stable contacts between CD4+ T cells and dendritic cells in vivo. J Exp Med 2006;203:505–511.
132 Mempel TR, Pittet MJ, Khazaie K, Weninger W, Weissleder R, von Boehmer H, von Andrian UH: Regulatory T cells reversibly suppress cytotoxic T cell function independent of effector differentiation. Immunity 2006;25:129–141.
133 Tang Q, Adams JY, Tooley AJ, Bi M, Fife BT, Serra P, Santamaria P, Locksley RM, Krummel MF, Bluestone JA: Visualizing regulatory T cell control of autoimmune responses in non-obese diabetic mice. Nat Immunol 2006;7:83–92.
134 Lim HW, Hillsamer P, Banham AH, Kim CH: Cutting edge: direct suppression of B cells by CD4+CD25+ regulatory T cells. J Immunol 2005; 175:4180–4183.
135 Tsuji M, Komatsu N, Kawamoto S, Suzuki K, Kanagawa O, Honjo T, Hori S, Fagarasan S: Preferential generation of follicular B helper T cells from FoxP3+ T cells in gut Peyer's patches. Science 2009;323:1488–1492.

Dr. Chang H. Kim
Laboratory of Immunology and Hematopoiesis, Department of Comparative Pathobiology
Purdue Cancer Center, Purdue University
725 Harrison Street, West Lafayette, IN 47907 (USA)
Tel. +1 765 494 0976, Fax +1 765 494 9830, E-Mail chkim@purdue.edu

Entschladen F, Zänker KS (eds): Cell Migration: Signalling and Mechanisms.
Transl Res Biomed. Basel, Karger, 2010, vol 2, pp 83–101

ADAMs and Ectodomain Proteolytic Shedding in Leukocyte and Tumour Cell Migration

Ann Ager[a] · Vera Knäuper[b] · Zara Poghosyan[c]

[a]Infection, Immunity and Biochemistry, and [c]Pathology, School of Medicine, and [b]Tissue Engineering, Dental School, Cardiff University, Cardiff, UK

Abstract

The ADAMs (a disintegrin and metalloproteinase) are a family of transmembrane proteins involved in 'ectodomain shedding' of adhesion molecules, chemokines, cytokines, cytokine receptors and growth factors. Metalloproteinase inhibitors, ADAM-deficient cells and cells expressing cleavage-resistant substrates have all been used to investigate the role of ADAMs in regulating cell adhesion and migration. In this chapter we will review the evidence for ADAM-dependent ectodomain shedding of cell adhesion molecules and chemokines in regulating leukocyte and tumour cell extravasation. Individual ADAMs are differentially regulated by alternative splicing, intracellular signalling pathways, cellular localization and availability of substrate. The regulation of ADAM10, 15 and 17 metalloproteinase activities during leukocyte extravasation and tumour cell metastasis will also be discussed.

The ADAMs (a disintegrin and metalloproteinase) are a family of multifunctional transmembrane proteins with key roles in cell-cell communication and ectodomain proteolysis (shedding) of a diverse array of cell surface receptors and ligands [1–3]. Proteolytically active ADAMs have been implicated in the regulation of cell adhesion and migration by virtue of their ability to cleave cell adhesion molecules (CAMs), chemokines, cytokines and cytokine receptors [4]. Following ADAM-dependent ectodomain cleavage, regulated intramembrane proteolysis of the intracellular domain by γ-secretase is crucial to downstream signalling of important developmental and disease-associated receptors such as Notch, CD44, E-cadherin and APP [5, 6]. ADAM expression is dysregulated in many cancers and individual ADAMs have been implicated in distinct stages of tumour progression [4, 7]. Tumour cell metastasis and leukocyte trafficking both involve the haematogenous spread of cells to secondary organs via the bloodstream. During this process, cells cross several physical barriers including epithelial linings, blood vessels, lymphatics and tissue stroma. ADAMs are regulators in negotiating these barriers [8, 9]. Here we will review the roles of ADAM-dependent ectodomain shedding in controlling leukocyte and tumour cell migration from the bloodstream into tissues. The mechanisms controlling ADAM metalloproteinase activity are still poorly understood but emerging data suggests that individual ADAMs are differentially regulated by signalling pathways and interactions both with substrates and other proteins. In this chapter we will focus on the regulation of ADAM10 and 17 metalloproteinase activity. We will also discuss the role of interacting kinases with cytoplasmic tails in regulating the metalloproteinase activity of ADAM15. The role of the disintegrin domain in regulating integrin-dependent cell adhesion and tumour cell-extracellular matrix (ECM) interactions has been reviewed recently [7].

Leukocyte Trafficking

The ability of the immune system to survey the body for invading microorganisms depends on the migration of leukocytes through solid organs and tissues via the bloodstream and lymphatics. Leukocytes are recruited to inflamed and injured tissues directly from the bloodstream and migrate through the interstitium to locate pathogen. Leukocyte migration into tissues involves a precisely coordinated sequence of adhesive interactions and signalling events in leukocytes and vascular endothelial cells [10]. It is a rapid process taking up to 10 min to complete and the ability to image the behaviour of leukocytes inside small blood vessels using intravital microscopy has allowed the molecular interactions to be dissected. Members of the selectin, integrin, immunoglobulin superfamily, CD44 and cadherin families of CAM are all involved in regulating leukocyte extravasation (also known as diapedesis). G-protein-coupled receptor activation by chemokines plays important roles in regulating CAM activity and guiding leukocyte migration. CAM engagement itself can trigger signalling in both the leukocyte and the ligand-expressing endothelial cell but the signalling events that regulate diapedesis are only just starting to be understood.

The type of organ and nature of the inflammatory insult will dictate the recruitment of distinct leukocyte subtypes via differential expression of CAMs on leukocytes and endothelial cells (EC) and local synthesis of chemokines by both the host and pathogen. In general, leukocytes initially *tether* and *roll* on the inner, endothelial surface of activated blood vessels using leukocyte (L-) or endothelial (E-, P-) selectins, and/or α_4 integrins (VLA4, $\alpha_4\beta_7$) in the case of mononuclear cells. Rolling leukocytes respond to chemokines immobilized on the EC surface which induce high affinity conformations in leukocyte integrins (LFA-1, Mac-1, VLA-4, $\alpha_4\beta_7$) resulting in *arrest* of rolling cells. Inside-out signalling following selectin and CD44 engagement also regulate integrin activation to stabilize adhesion. Arrested leukocytes spread, polarize and *crawl* on the luminal blood vessel surface in search of exit points. Leukocytes *transmigrate* the endothelial lining by penetrating the apical surface and exiting the basolateral surface.

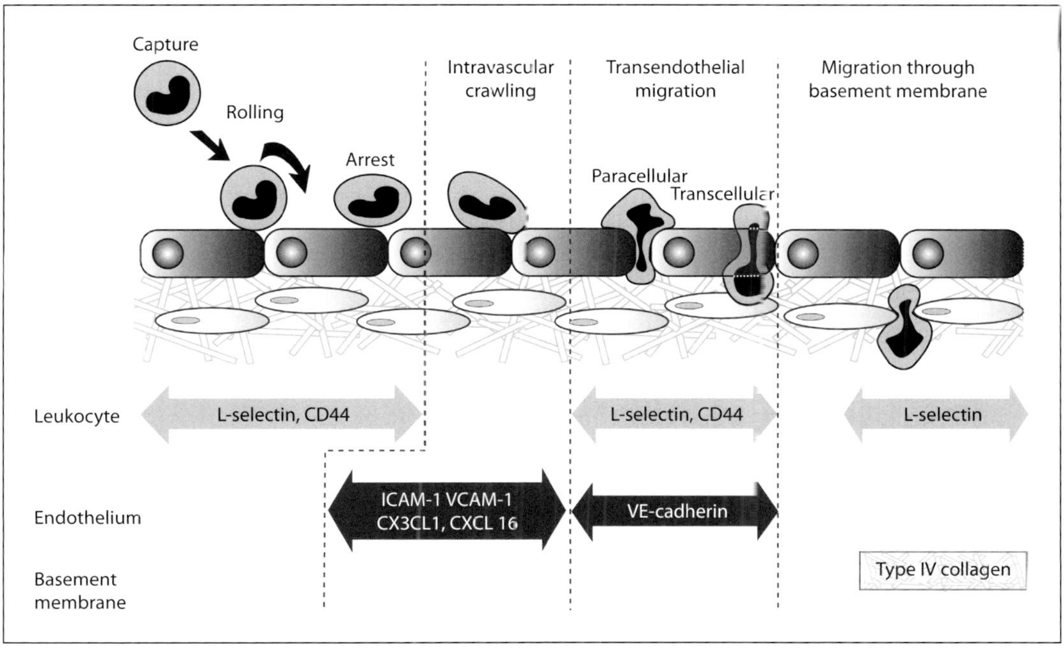

Fig. 1. Ectodomain shedding of receptors on leukocytes and vascular EC. Leukocyte recruitment from the bloodstream into lymphoid organs and sites of inflammation is regulated by a coordinated sequence of adhesive interactions which can be separated into distinct stages. CAMs and chemokines that have been shown to undergo ADAM-dependent ectodomain shedding from leukocytes and EC are shown and their demonstrated or proposed roles in regulating different stages of leukocyte recruitment outlined.

The predominant route is between adjacent EC and leukocytes engage endothelial junctional proteins such as CD31, CD99 and the junctional adhesion molecules (JAMs) to drive paracellular migration. Leukocytes also pass transcellularly through the cytoplasm of single EC using F-actin-containing podosomes for invasion [11]; the impact on leukocyte and endothelial function is likely to differ depending on the route of transendothelial migration. Finally, leukocytes penetrate the surrounding basement membrane and embedded sheath of pericytes to enter tissue parenchyma (fig. 1).

An important feature of this cascade is that it is tightly regulated both spatially and temporally such that leukocyte recruitment is rapid and efficient and the integrity of vessel wall is not impaired. For example, leukocytes roll on endothelium at appropriate speeds to be able to respond to activating ligands on the EC surface. Contact time between leukocytes and EC during arrest and transmigration should be sufficient to avoid premature activation of leukocyte or EC. A second important feature is that recruitment should be rapidly abrogated during resolving inflammation and vessels returned to a preinflamed state. This is facilitated by each step of the cascade being independently regulated and reversible. For example, rolling and adherent leukocytes can be released back into the circulation rather than undergo transendothelial migration. Recent studies have found that transendothelial migration is also reversible, allowing leukocytes to regain access to the circulation if entry into the interstitial compartment is inhibited or the barrier function of endothelium is impaired [12, 13]. Leukocytes which have been recruited from the bloodstream and do not apoptose usually exit tissues via the lymphatics. However, it is also possible that leukocytes directly re-enter the circulation by intravasation, i.e., reverse transmigration of the vessel wall, as is proposed for metastatic tumour cells entering the circulation. Understanding the mechanisms that regulate each stage of leukocyte migration is important for the development of anti-inflammatory therapeutics. Parallel studies of tumour metastasis may reveal overlapping or shared mechanisms controlling tumour cell entry into, or exit from the circulation, which could be applicable to cancer therapy.

Several approaches, including the use of broad-spectrum metalloproteinase inhibitors with differing selectivities for ADAMs vs. matrix metalloproteinases (MMPs) and ADAM-deficient cells and mice, have demonstrated that ectodomain shedding of receptors on leukocytes and EC regulate different stages of leukocyte recruitment (fig. 1). A single ADAM cleaves multiple substrates; additional approaches will, therefore, be required to dissect shedding of individual CAMs/chemokines/cytokines in leukocyte recruitment. The development of genetically modified mice expressing cleavage-resistant CAMs is proving useful in this analysis [14–16].

ADAMs and Ectodomain Shedding of Leukocyte Receptors

L-Selectin/CD62L

L-selectin mediates the initial capture (tethering) and rolling of leukocytes in inflamed blood vessels and of lymphocytes in lymph node high endothelial venules (HEV). L-selectin-mediated adhesion needs to resist the disruptive forces of blood flow and this is achieved by anchoring of the cytoplasmic tail to the cortical actin cytoskeleton [17]. L-selectin is spatially distributed to the tips of microvilli via ezrin-radixin-moesin (ERM) proteins interacting with its cytoplasmic tail, and this is required for efficient capture from flowing blood [18, 19]. Clustering of L-selectin at the leukocyte surface by its multivalent ligands activates β_1 and β_2 subunit-containing integrins [20–22] and polarization [23] which promote adhesion strengthening and transendothelial migration [24–26]. L-selectin-mediated inside-out signalling also stimulates rapid mobilization of CXCR4 chemokine receptor to the leukocyte surface [27] which may control CXCL12-dependent transendothelial migration [28, 29].

Regulation of L-Selectin Shedding

L-selectin was one of the first CAMs shown to undergo ADAM-dependent ectodomain shedding. Shedding reduces the cell surface density of L-selectin and generates

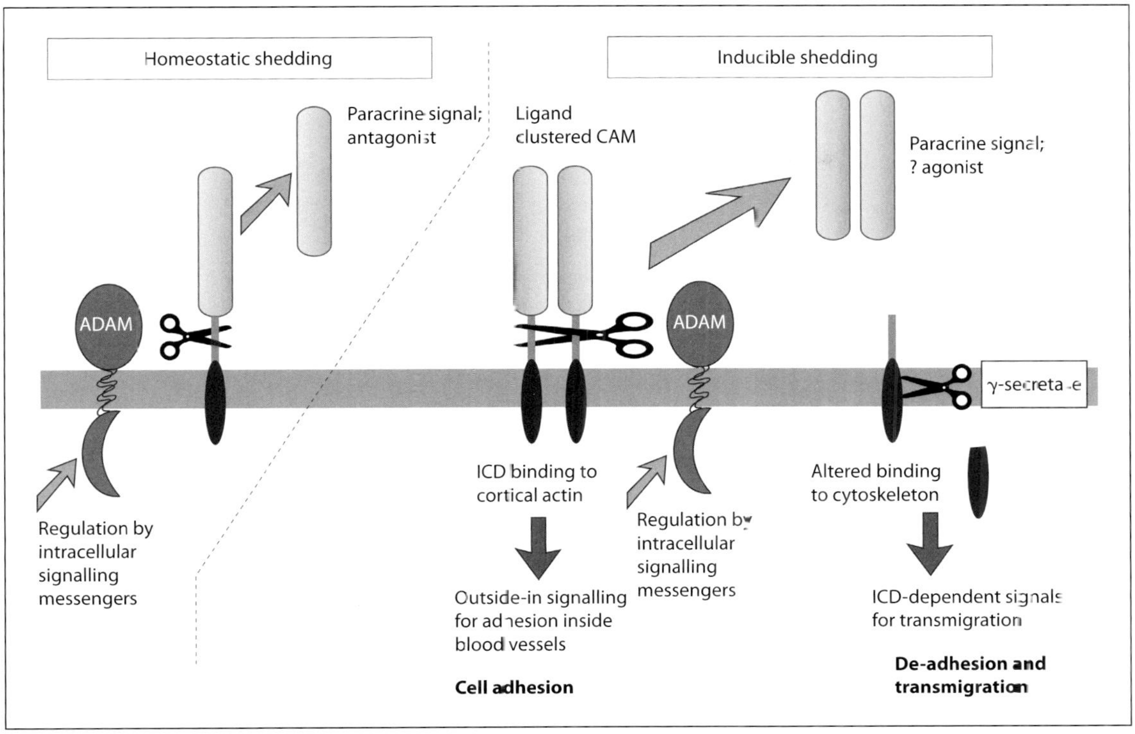

Fig. 2. Regulation of leukocyte recruitment by ADAM-dependent cleavage of CAMs. Constitutive or homeostatic shedding of CAMs generates a soluble, monomeric ectodomain which functions in a paracrine manner as antagonist. Ligand-clustered CAM intracellular domain (ICD) binds to cortical actin cytoskeleton via ERM proteins to stabilize leukocyte-endothelial cell adhesion in flowing blood. Activation-induced ectodomain shedding cleaves ligand-clustered CAM to release polymeric ectodomains with potential agonist properties. Release of the ectodomain abrogates ICD linkages to actin cytoskeleton to facilitate de-adhesion and transmigration. Regulated intramembrane proteolysis of ICD may be required for de-adhesion and transmigration. Homeostatic and inducible shedding will be regulated by different ADAMs in different cells.

a monomeric, soluble species containing the N-terminal lectin domain which competes with cell surface L-selectin. A single ADAM-dependent cleavage site in the 15-amino-acid membrane proximal region (MPR) of L-selectin has been mapped. Truncation of the MPR by 8 or 9 residues or substitution with the shorter homologous domains from E- or P-selectin generated mutants of L-selectin which resist phorbol ester-induced shedding [30–33]. However, these mutants are not completely null for L-selectin turnover since basal and ligand-induced shedding are detectable to varying levels [14, 15, 32]. In addition, some MPR mutants show altered L-selectin-dependent adhesion suggesting that signalling is compromised [15].

In the absence of overt stimulation, isolated leukocytes constitutively shed L-selectin in vitro which may reflect turnover in the membrane (fig. 2). The rate of shedding is substantially accelerated by a range of activators including G-protein-coupled receptors [34], T-cell receptor [35], cross-linking using immobilized antibodies [32], non-steroidal anti-inflammatory drugs [36], annexin 1 [37], HIV [38] and pharmacological manipulation of PKC and intracellular calcium levels using phorbol esters and ionophores respectively [39]. Early studies showed that calmodulin binding to the membrane proximal cytoplasmic tail of L-selectin negatively regulated shedding and calmodulin antagonists facilitated L-selectin shedding [40]. L-selectin shedding induced by phorbol esters and non-steroidal anti-inflammatory drugs is regulated by ADAM17 [41, 42], whereas calcium flux and P2X7 receptor engagement activate ADAM10 [39]. Calmodulin antagonists also activate ADAM10 [43] and ADAM8 regulates activation-dependent shedding of L-selectin from myeloid cells [44]. Interestingly, shedding of L-selectin induced by antibody cross-linking, which is thought to mimic ligand engagement, is resistant to MPR truncation and is not susceptible to hydroxamate (TAPI) inhibition suggesting that it may be metalloproteinase independent [45].

Cell surface levels of L-selectin on leukocytes in lethally irradiated mice reconstituted with ADAM17-deficient stem cells are 3- to 5-fold higher than on wild-type controls. ADAM17-deficient leukocytes failed to cleave L-selectin in response to PMA and leukocytes recruited to sites of inflammation did not shed L-selectin [46]. Circulating levels of soluble L-selectin are reduced and cell surface expression increased in mice expressing cleavage-resistant L-selectin suggesting a reciprocal relationship [14, 15]. However, the fact that leukocyte-expressed ADAM17 regulates cell surface and not soluble L-selectin suggests that the two are independently regulated [46]. Lymphoid and myeloid cells both contribute soluble L-selectin [47] and basal turnover, mechanical or ligand-induced shedding during leukocyte rolling and shedding from apoptosing leukocytes have all been suggested as possible sources [48, 49].

Regulation of Leukocyte Recruitment by L-Selectin Shedding

The levels of circulating soluble L-selectin at 2–10 μg/ml (~30–150 nM) are sufficient to inhibit L-selectin-dependent recruitment of lymphocytes into lymph nodes [47], even though the affinity of monomeric L-selectin is low [50]. It has been proposed that soluble L-selectin limits excess leukocyte recruitment during ongoing inflammation, and there is some evidence to support this hypothesis [51]. L-selectin is rapidly downregulated from the cell surface on lymphocytes during transmigration of lymph node HEV [52, 53] and leukocytes harvested from sites of inflammation express low levels of L-selectin [34] but what role, if any, L-selectin shedding from the leukocyte surface plays in recruitment was undetermined. Early studies showed clearly that shedding from the leukocyte surface was not required for leukocytes to undergo transendothelial migration. However, by controlling cell surface levels of L-selectin, shedding provided a negative feedback mechanism to limit excess neutrophil recruitment. Leukocytes unable to shed L-selectin showed altered recruitment into tissues, but the effects on neutrophils and T-lymphocytes vary depending on the inflammatory stimulus and site of recruitment.

Leukocyte Rolling and Recruitment Inside Blood Vessels
Administration of broad-spectrum metalloproteinase inhibitors reduced the velocity of neutrophil rolling in mice or on immobilized L-selectin ligands in vitro [54, 55]. The rolling velocity is dependent on the density of L-selectin at the leukocyte surface [56]; the reduced rolling velocity in the presence of shedding inhibitors may therefore reflect increased L-selectin expression following blockade of homeostatic or adhesion-induced shedding, although this was not measured directly. A further study in a model of cytokine-dependent inflammation showed increased LFA-1-dependent adhesion and spreading of neutrophils inside inflamed vessels of mice when L-selectin shedding was blocked using pharmacological inhibitors [57]. However, genetically modified knock-in mice in which leukocytes express a mutant form of L-selectin that resists shedding (L/E-selectin mice), showed normal rolling and recruitment of neutrophils in cytokine-dependent inflammation [15]. These data suggest that metalloproteinase inhibitors regulate neutrophil recruitment by blocking shedding of substrates in addition to L-selectin; however, the effects metalloproteinase inhibitor on plasma levels of soluble L-selectin and on L-selectin-dependent signalling, which are both reduced in L/E-selectin mice, were not determined, which may account for differences between pharmacologic and genetic inhibition of L-selectin shedding [15, 57]. The impact of L-selectin shedding on neutrophil recruitment does depend on the nature of the inflammatory stimulus since L-selectin shedding had little or no impact on the slow rolling or recruitment of neutrophils on E-selectin expressing endothelium, presumably because E-selectin-mediated signalling in leukocytes dominates over L-selectin-dependent signalling [54, 58].

Pharmacological inhibition of L-selectin shedding did not control L-selectin-dependent rolling of T lymphocytes on immobilized ligands in vitro or the recruitment of lymphocytes in lymph node HEV [14, 52]. T cells from transgenic mice expressing a shedding-resistant form of L-selectin in which the MPR was substituted with that of P-selectin (L/P-selectin mice) were also recruited normally following intravenous administration [14]. These results suggest that fluctuations in cell surface L-selectin during tethering and rolling, if they do occur, do not regulate recruitment of T cells into lymph nodes. The level of L-selectin on naive and memory T cells is saturating for lymph node entry since T cells overexpressing L-selectin do not show increased recruitment [56]. However, higher expression of L-selectin by CD4+CD25+ regulatory T cells in L/E-selectin mice significantly increases recruitment into lymph nodes [59]. This may reflect the more important role of L-selectin-dependent signalling during recruitment of T-regulatory cells over conventional T cells. Lymphocytes expressing low levels of L-selectin, such as antigen-activated T cells, B lymphocytes and CLL cells, will roll faster and have a reduced contact time with HEV; recruitment may, therefore, be sensitive to fluctuations in L-selectin expression due to shedding [56, 60, 61]. Fluctuations in L-selectin expression due to shedding may also have more impact on lymphocyte recruitment to sites of inflammation where ligand expression is lower than in HEV.

Leukocyte Transendothelial Migration

L-selectin shedding also regulates leukocyte transmigration across the vessel wall but, as described above, the effect of inhibiting shedding differs between T cells and neutrophils and varies depending on the tissue and inflammatory stimulus. T cells from L/P-selectin mice expressing non-cleavable L-selectin transmigrated HEV at a slower rate than T cells expressing wild-type L-selectin and accumulated in the vessel wall and the parenchyma surrounding HEV [14]. In another study, T cells expressing L/E-selectin transmigrated HEV and entered the lymph node parenchyma but the rate of transmigration was not compared directly with wild-type T cells [15]. Systemic treatment of mice with selective metalloproteinase inhibitors that block L-selectin shedding resulted in accumulation of lymphocytes within the endothelial cell lining of HEV and reduced exit at the basolateral surface [52]. Since T cells expressing cleavage-resistant L-selectin exited the basolateral surface and did not preferentially accumulate within the walls of HEV, other metalloproteinase-dependent substrates may control exit from the basolateral surface of endothelium such as CD31 or VE-cadherin (see later) at inter-endothelial junctions. Alternatively, the low residual turnover of mutant L/P-selectin may regulate exit of T cells from the basolateral surface of HEV in L/P-selectin mice. Exogenously applied chemokines stimulated normal rolling and adhesion of neutrophils in L/E-selectin mice, however, L-selectin-dependent extravasation and migration into the interstitium [62] were severely impaired in L/E-selectin cleavage-resistant mice [15]. In contrast, neutrophil extravasation in cytokine-dependent inflammation was not altered whereas recruitment to inflamed peritoneum was increased in L/E-selectin shedding resistant mice, but whether this was due to altered rolling, adhesion or transmigration was not determined.

L-selectin-dependent activation of integrins and chemokine receptor mobilization may be required to stabilize rolling, arrest and/or polarization of leukocytes inside blood vessels. However, L-selectin engagement alters its anchorage to the actin cytoskeleton and prevents F-actin polymerization in response to ICAM-1 engagement [63], which has been implicated in podosome-dependent transcellular migration of lymphocytes [11]. It is therefore possible that ectodomain shedding of L-selectin is required to terminate or modify downstream signalling and allow leukocytes to progress to the next stage of diapedesis (fig. 2). Ligands for L-selectin are expressed both inside and outside of blood vessels [64] and this may explain the pleiotropic effects of inhibiting L-selectin shedding. Chronic L-selectin signalling induced by ligands expressed on the apical surface of endothelium may abrogate intraluminal crawling; ligands expressed on the apical and basolateral endothelial surfaces, such as in HEV, may retard lymphocyte transmigration and ligands expressed on abluminal surface and/or in basement membrane will result in perivascular cuffing.

CD44 Shedding and T-Cell Recruitment in Autoimmune Lesions

The role of L-selectin shedding in regulating T-lymphocyte transendothelial migration into lymph nodes is remarkably similar to that described for MMP-dependent shedding of CD44 controlling infiltration of diabetogenic T cells into pancreatic islets [65]. Global shedding of CD44 by exogenous treatment with membrane type (MT) 1-MMP resulted in lack of recruitment via endothelial expressed hyaluronan due to loss of homing molecule expression on T cells. Systemic treatment of mice with a hydroxamate-based MMP inhibitor reduced infiltration of the islet, although the metalloproteinase responsible was not identified and several CD44 sheddases (including ADAM10 and 17) [43] are possible targets in vivo. Since ectodomain shedding of CD44 was stimulated in insulin-specific CD8 T cells following adhesion to matrix, the authors hypothesize that stimulated shedding of CD44 following adhesion to the inner endothelial is required for diapedesis. CD44 is anchored to the cortical cytoskeleton via ERM proteins and, as suggested for L-selectin, ectodomain shedding of ligand-clustered CD44 may be required to promote de-adhesion and transendothelial migration (fig. 2). T cells expressing cleavage-resistant forms of CD44 will be useful to test this hypothesis.

ADAMs and Ectodomain Shedding of Endothelial Receptors

ADAM-regulated proteolysis of receptors on EC has also been shown to control leukocyte adhesion and transendothelial migration (fig. 1). For example, the membrane-inserted chemokines CX3CL1 (frackalkine) and CXCL16 are both cleaved from human umbilical vein endothelial cells (HUVECs) by endogenous ADAM10 and cleavage leads to detachment of adherent leukocytes [66, 67]. ADAM17 additionally cleaves CX3CL1 [68]. Endogenous ADAM10 has also been shown to cleave the endothelial junctional protein VE-cadherin [69] which undergoes homophilic interactions with adjacent EC to maintain integrity of the endothelial lining [70]. Inhibition of ADAM10 in HUVECs as well as in T cells partially blocks transmigration [69] but whether VE-cadherin shedding is required for transmigration was not demonstrated directly. ADAM10-dependent shedding of VE-cadherin and membrane-inserted chemokines is stimulated by calcium ionophore [67] and thrombin stimulates ADAM10-mediated VE-cadherin proteolysis, possibly via increased calcium flux [69]. Leukocyte adhesion is associated with an increase of intracellular Ca^{2+} [71], which may also regulate ADAM10-dependent shedding of VE-cadherin and membrane chemokines to facilitate leukocyte de-adhesion and transmigration.

Endothelial ligands for leukocyte integrins ICAM-1 and VCAM-1 also undergo activation-dependent ectodomain shedding in vitro which has been shown to be ADAM17 dependent following cytokine or PKC activation [72, 73]. In response to leukocyte adhesion, ICAM-1 and VCAM-1 become clustered via ERM protein binding to actin cytoskeleton in so-called transmigratory cups on the apical surface of EC [74, 75]. Cleavage of ICAM-1 or VCAM-1 may be required for leukocytes to move on the apical surface to locate exit points for transmigration or to regulate signalling in EC required for transendothelial migration (fig. 2).

In a separate study of lymphocyte transendothelial migration in vitro, engagement of VCAM-1 on the endothelial surface stimulates lymphocyte transmigration which is accompanied by endothelial NADPH oxidase generation of ROS and activation of MMP9. Pretreatment of EC, but not lymphocytes, with a broad-spectrum hydroxamate ADAM/MMP inhibitor GM6001 blocked lymphocyte transmigration, but the metalloproteinases and substrates responsible have not yet been identified [76].

Crossing the Basement Membrane and Interstitial Migration

Although ADAMs can degrade some ECM proteins, members of the related MMP family have been implicated in penetration of basement membranes. In vitro models of T-cell invasion have shown that binding to VCAM-1 stimulates invasion of basement membrane collagen via induction of MMP9 expression, although this is a slow response taking several hours [77]. This may be important for penetration of the tightly regulated endothelial cell layers, such as in the blood-brain barrier. Studies using TIMP1 and synthetic broad-spectrum MMP inhibitors which do not block ADAMs did not affect T-cell migration across blood vessel basement membranes of mouse lymph nodes or spleen [52]. Recent studies have shown that neutrophils penetrate areas of low matrix density within basement membranes of inflamed vessels [78]; pericellular proteolysis may therefore not be required for leukocytes to cross basement membranes.

Once diapedesis is complete, the migration of lymphocytes within tissues is controlled by chemokines, but whether this is regulated by ADAM-dependent pericellular proteolysis has not been addressed. Soluble MMPs secreted into the interstitium establish chemokine gradients and generate ECM fragments that control migration [79]. Systemic treatment of mice with TIMP1 and synthetic broad-spectrum MMP inhibitors did not block T-cell movement into the interstitium of lymphoid organs [52]. This is not surprising since TIMP1 does not block ADAMs or the membrane-inserted MT-MMPs. Migrating lymphocytes and dendritic cells entering lymph nodes are not exposed to ECM proteins, instead crawling along the reticular cells of the lymph nodes [80]. The architecture and composition of stroma varies significantly between tissues and leukocyte migration in non-lymphoid organs may be regulated by metalloproteinases, as shown for tumour cells migrating through 3D extracellular matrix [81], although this will depend on the precise composition and structure of stromal compartments within different tissues [82].

ADAMs and Ectodomain Shedding in Tumour Cell Extravasation

In experimental animal models, extravasation of metastatic tumour cells administered directly into the bloodstream is markedly slower than that of leukocytes, taking up to 24 h to complete [83]. It is currently unclear whether tumour cells diapedese vascular endothelium in the same way as leukocytes. Adhesion molecules and chemokines that regulate metastasis are just starting to be identified. Glycoconjugates at the tumour cell surface binding to endothelial selectins (E- and P-selectin) have been implicated in metastasis [84] and those modified by $\alpha_{2,6}$-sialotransferase ST6GALNAC5 in the metastasis of breast cancer cells to the brain [85]. Tumour cell metastasis is facilitated by P-selectin-mediated interactions with platelets and L-selectin-dependent interactions of myeloid cells either with tumour cells or with induced ligands on involved endothelium [86, 87]. Studies in which metastatic tumour cells interacting with blood vessels in the lung have been imaged directly rarely found extravasated tumour cells and concluded that diapedesis was not a common occurrence. Instead it was proposed that endothelium-attached tumour cells proliferate intravascularly and, with time, metastatic foci outgrow and destroy the vessel wall [88]. A future challenge in the field will be to identify the molecular basis of spontaneous tumour metastasis.

The role of ADAMs in tumour metastasis is just starting to be dissected. Since ectodomain shedding of CD44 via ADAM10 or ADAM17 regulates the migration of tumour cells on ECM [43] and CD44 shedding has been implicated in regulating inflammatory T-cell extravasation [65], it is tempting to speculate that a similar pathway might operate during tumour extravasation. Early studies using broad-spectrum MMP/ADAMs inhibitors did not inhibit extravasation of administered metastatic tumour cells [89]. However, ADAM15 supports cellular adhesion and motility in several systems and binds to endothelial integrins via an RGD motif contained in its disintegrin domain. Loss of ADAM15 in prostate cancer PC3 cells attenuated the migratory capacity in a wound healing assay and decreased PC3 basal adhesion to fibronectin, laminin, and vitronectin [90]. Downregulation of ADAM15 in PC3 cells resulted in significant reduction in endothelial adhesion, as well as in transendothelial migration of PC3 cells through HUVEC and HDMEC monolayers [91]. The interaction between ADAM15 and specific integrins or other transmembrane receptors may coordinate the proteolytic activity of ADAM15 directly to regulate adhesion and transendothelial migration, but this has not yet been tested.

How, Where and When Are ADAMs Activated

The precise spatio-temporal regulation of the adhesion cascade means that shedding events must be tightly regulated to ensure progression of rolling cells to arrest rather than detachment back to the circulation and progression of arrested cells to transmigration without breaching the vessel wall. In vivo studies have shown that global shedding of L-selectin from the leukocyte surface does not occur during rolling or binding to EC inside blood vessels [92]. It is unlikely that global shedding of endothelial receptors takes place, since it would result in detachment of leukocytes or breaching of the endothelial barrier. Activation of ADAMs at the leukocyte and/or endothelial cell surface during recruitment is, therefore, likely to be restricted to particular domains in apposing membranes. ADAMs generally cleave substrates in the same membrane (i.e., in *cis*). An important control point is, therefore, how ADAMs and substrate come together in the membrane. However a recent report showed ADAMs cleavage of substrate on opposing cells (i.e., in *trans*) during heterotypic cellular interactions [93]. It is therefore possible that during recruitment, ADAMs expressed at the leukocyte surface cleave endothelial receptors and vice versa. Further studies will be required to determine how and where ADAM17 and ADAM10 are activated in leukocytes and EC during binding and transmigration and whether additional ADAMs regulate leukocyte extravasation.

Regulation of ADAM10 and 17 Proteolytic Activity

ADAM17 is widely expressed and is reported to cleave a large number of substrates including CAMs and chemokines on leukocytes and EC [4]. It plays a key role in processing ligands for epidermal growth factor receptor and is essential during embryonic development [2, 41]. A number of studies have demonstrated the importance of PMA-PKC-Erk-ADAM17 activation cascade in the processing of various transmembrane proteins. In response to PMA, Erk phosphorylates the cytoplasmic domain of ADAM17 at Thr735, leading to the translocation of ADAM17 to the cell surface [94], potentially regulating the proteolytic activity of the metalloproteinase domain. In antigen-activated T cells, PI3K p110δ subunit activation of erk kinase may also regulate L-selectin shedding [95] via phosphorylation of Thr735 in ADAM17. However, Horiuchi et al. [96] demonstrated that neither the transmembrane nor the cytoplasmic domain of ADAM17 are required for PMA-induced shedding of EGFR ligands, but the intact ectodomain of ADAM17 is indispensible for PMA-induced shedding. This could either be because of the necessity of the ectodomain to interact with other regulatory membrane-anchored accessory proteins, or because the intact ectodomain is necessary for proper presentation of the catalytic domain and/or membrane localization.

Extra- and intracellular domains in individual substrates also regulate ADAM17-dependent shedding, but there are no set rules for all substrates. For example, L-selectin shedding is regulated by the EGF-containing ectodomain [97]. Complete removal of the cytoplasmic tail reduced shedding by ~50% whereas deletion of carboxy-terminal 11 amino acids from the 17-amino-acid tail had no effect on PMA-induced shedding [30]. Calmodulin binding to a site which overlaps with that of ERM proteins binding inhibits ADAM17-dependent shedding [40] and substitution of ERM-binding sites within the membrane proximal cytoplasmic tail of L-selectin also abrogates shedding

[98]. In contrast, the cleavage site of the EGFR ligand, TGF-α is necessary and sufficient for ADAM17-dependent processing [96].

The cytoplasmic domain of ADAM10 contains a sequence resembling the calcium-dependent IQ consensus binding site for calmodulin (IQXXXRXXXXR). This sequence has a potential role in ADAM10 maturation, and/or stabilization of the mature protein. In resting cells, pro-ADAM10 is constitutively associated with calmodulin, and an increase in intracellular calcium would activate ADAM10 by promoting the dissociation of calmodulin [43]. However, Horiuchi et al. [96] demonstrated that ADAM10IQ>AA mutation did not affect the ability of this mutant to rescue ionomycin-stimulated shedding of EGRF ligands compared to wild-type ADAM10 in ADAM10$^{-/-}$ cells, suggesting that other sequences in the cytoplasmic domain regulate the response of ADAM10 to calcium influx. In support of this, the removal of the cytoplasmic domain strongly decreased its response to ionomycin.

Calcium flux-induced ectodomain cleavage of CD44 is mediated by ADAM10, by promoting ADAM10 dissociation from calmodulin [43]. The same authors showed that phorbol ester-stimulated ADAM17-mediated CD44 shedding is dependent on Rac-GTPase, which is necessary for cytoskeletal rearrangements and ADAM17 membrane localization. Murai et al. [99] demonstrated that EGF-regulated cleavage of CD44 is also mediated by ADAM10 via activation of Rac GTPase, where EGF-induced erk activation was also necessary for CD44 cleavage. A potential mechanism would be Rac GTPase – Pak1 – Raf1 – erk – ADAM10 phosphorylation and activation. Thus, both ADAM10 and ADAM17, differentially regulated by Ca^{2+} influx and PKC, inhibit CD44-dependent adhesion to EC or ECM adhesion. Interestingly, CD44 ectodomain cleavage, independently mediated by ADAM10 and ADAM17, are both required for efficient migration of tumour cells on immobilized ligand [43].

Regulation of ADAM15 Function by Interacting Kinases

Receptor or non-receptor kinases are emerging as regulators of ADAM function. In this respect, the C-terminal intracellular domain of ADAM proteins can provide insight into potential mechanisms of such regulation. The C-terminal cytoplasmic domain varies both in size and sequence in different ADAMs. In most proteolytically active ADAMs this domain contains multiple proline-rich regions and/or phosphorylation sites such as tyrosines, enabling them to interact with SH3 and SH2 domain-containing proteins [7]. These intracellular interactions with signalling molecules may modulate ADAM functions, such as ADAM-mediated proteolytic processing of membrane-bound proteins. On the other hand, due to these cytoplasmic domain-mediated protein-protein interactions, ADAMs are becoming important factors in intracellular signalling. To this end, of particular interest is ADAM15. The intracellular domain of ADAM15 contains multiple proline-rich motifs and tyrosine residues that mediate association with cytoplasmic signalling molecules, such as non-receptor tyrosine kinases and adaptor molecules [100]. Importantly, ADAM15 is subject to complex alternative splicing that leads to differential inclusion of these SH2 and SH3 ligands, and we have recently shown that the expression levels of particular splice variants is linked to prognosis in human breast cancers [101]. Alternative splicing affects only the intracellular domain of ADAM15, not the extracellular modules, indicating it may be important in the regulation of tumour cell behaviour. Of particular interest in this context is the discovery that the ADAM15B variant (but not ADAM15A) associates with Src, resulting in ADAM15B phosphorylation, which upregulates the proteolytic activity of the extracellular domain, leading to enhanced cleavage of fibroblast growth factor receptor 2iiib (FGFR2iiib) [102]. The overexpression of individual ADAM15 variants in breast cancer cells has profoundly opposing effects on cell phenotype and morphology, with adhesion, migration and invasion enhanced by ADAM15A, and reduced adhesion and motility by ADAM15B [101]. These cells also have significantly different cytoskeletal architecture. ADAM15A-expressing cells are well spread and have prominent actin stress fibres, while ADAM15B-expressing cells are rounded, less spread, with insignificant filamentous actin and strong cortical actin, suggesting that ADAM15 alternative splicing affects actin reorganization, hence cellular motility and invasive properties of these cells [101]. These data support the concept of 'inside-out' signalling control, whereby ADAM15/kinase association via ICD motifs regulates the proteolytic activity and function of ADAM15 variants in breast cancer.

Conclusions and Future Prospects

There is still much to learn about the specific roles of ADAM-dependent ectodomain shedding in leukocyte recruitment and in tumour progression. However, studies using animal models of disease have highlighted the potential of L-selectin cleavage as a target for anti-inflammatory therapies. Global shedding of L-selectin from neutrophils in the vascular lumen limits leukocyte recruitment and this may be a contributory mechanism of some non-steroidal anti-inflammatory drugs. A desirable target may be to induce L-selectin cleavage in the absence of leukocyte activation to limit infiltration during ongoing inflammation. However, inhibition of L-selectin cleavage may have differential outcomes depending on the type of inflammatory disease [15, 103] and therefore may not always be a suitable target.

Future studies using in vitro models of extravasation and animal models of metastasis are required to identify potential ADAMs and relevant substrates for therapeutic intervention. Once identified, the development of selective inhibitors that target individual ADAMs and/or individual substrates will be required to avoid unwanted side effects on normal developmentally regulated pathways in which ADAMs proteolysis are also involved [4].

References

1 White JM: ADAMs: modulators of cell-cell and cell-matrix interactions. Curr Opin Cell Biol 2003;15: 598–606.

2 Blobel CP: ADAMs: key components in EGFR signalling and development. Nat Rev Mol Cell Biol 2005;6:32–43.

3 Reiss K, Saftig P: The 'a disintegrin and metalloprotease' (ADAM) family of sheddases: physiological and cellular functions. Semin Cell Dev Biol 2009;20: 126–137.

4 Murphy G: The ADAMs: signalling scissors in the tumour microenvironment. Nat Rev Cancer 2008;8: 929–941.

5 Selkoe DJ, Wolfe MS: Presenilin: running with scissors in the membrane. Cell 2007;131:215–221.

6 Okamoto I, Kawano Y, Murakami D, Sasayama T, Araki N, Miki T, Wong AJ, Saya H: Proteolytic release of CD44 intracellular domain and its role in the CD44 signaling pathway. J Cell Biol 2001;155: 755–762.

7 Edwards DR, Handsley MM, Pennington CJ: The ADAM metalloproteinases. Mol Aspects Med 2008; 29:258–289.

8 Reiss K, Ludwig A, Saftig P: Breaking up the tie: disintegrin-like metalloproteinases as regulators of cell migration in inflammation and invasion. Pharmacol Ther 2006;111:985–1006.

9 Garton KJ, Gough PJ, Raines EW: Emerging roles for ectodomain shedding in the regulation of inflammatory responses. J Leukoc Biol 2006;79: 1105–1116.

10 Ley, K, Laudanna, C, Cybulsky, MI, Nourshargh, S: Getting to the site of inflammation: the leukocyte adhesion cascade updated. Nat Rev Immunol 2007; 7:678–689.

11 Carman CV, Sage PT, Sciuto TE, de la Fuente MA, Geha RS, Ochs HD, Dvorak HF, Dvorak AM, Springer TA: Transcellular diapedesis is initiated by invasive podosomes. Immunity 2007;26:784–797.

12 Bradfield PF, Scheiermann C, Nourshargh S, Ody C, Luscinskas FW, Rainger GE, Nash GB, Miljkovic-Licina M, Aurrand-Lions M, Imhof BA: JAM-C regulates unidirectional monocyte transendothelial migration in inflammation. Blood 2007;110:2545–2555.

13 McGettrick HM, Hunter K, Moss PA, Buckley CD, Rainger GE, Nash GB: Direct observations of the kinetics of migrating T cells suggest active retention by endothelial cells with continual bidirectional migration. J Leukoc Biol 2009;85:98–107.

14 Galkina E, Tanousis K, Preece G, Tolaini M, Kioussis D, Florey O, Haskard DO, Tedder TF, Ager A: L-selectin shedding does not regulate constitutive T cell trafficking but controls the migration pathways of antigen-activated T lymphocytes. J Exp Med 2003; 198:1323–1335.

15 Venturi GM, Tu L, Kadono T, Khan AI, Fujimoto Y, Oshel P, Bock CB, Miller AS, Albrecht RM, Kubes P, Steeber DA, Tedder TF: Leukocyte migration is regulated by L-selectin endoproteolytic release. Immunity. 2003;19:713–724.

16 Wirth TC, Badovinac VP, Zhao L, Dailey MO, Harty JT: Differentiation of central memory CD8 T cells is independent of CD62L-mediated trafficking to lymph nodes. J Immunol 2009;182:6195–6206.

17 Kansas GS, Ley K, Munro JM, Tedder TF: Regulation of leukocyte rolling and adhesion to high endothelial venules through the cytoplasmic domain of L-selectin. J Exp Med 1993;177:833–838.

18 Von Andrian UH, Hasslen SR, Nelson RD, Erlandsen SL, Butcher EC: A central role for microvillous receptor presentation in leukocyte adhesion under flow. Cell 1995;82:989–999.

19 Ivetic A, Florey O, Deka J, Haskard DO, Ager A, Ridley AJ: Mutagenesis of the ezrin-radixin-moesin binding domain of L-selectin tail affects shedding, microvillar positioning, and leukocyte tethering. J Biol Chem 2004;279:33263–33272.

20 Hwang ST, Singer MS, Giblin PA, Yednock TA, Bacon KB, Simon SI, Rosen SD: GlyCAM-1, a physiologic ligand for L-selectin, activates β_2 integrins on naive peripheral lymphocytes. J Exp Med 1996; 184:1343–1348.

21 Giblin PA, Hwang ST, Katsumoto TR, Rosen SD: Ligation of L-selectin on T lymphocytes activates β_1 integrins and promotes adhesion to fibronectin. J Immunol 1997;159:3498–3507.

22 Simon SI, Cherapanov V, Nadra I, Waddell TK, Seo SM, Wang Q, Doerschuk CM, Downey GP: Signaling functions of L-selectin in neutrophils: alterations in the cytoskeleton and colocalization with CD18. J Immunol 1999;163:2891–2901.

23 Harris H, Miyasaka M: Reversible stimulation of lymphocyte motility by cultured high endothelial cells: mediation by L-selectin. Immunology 1995;84: 47–54.

24 Shamri R, Grabovsky V, Gauguet JM, Feigelson S, Manevich E, Kolanus W, Robinson MK, Staunton DE, von Andrian UH, Alon R: Lymphocyte arrest requires instantaneous induction of an extended LFA-1 conformation mediated by endothelium-bound chemokines. Nat Immunol 2005;6:497–506.

25 Phillipson M, Heit B, Colarusso P, Liu L, Ballantyne CM, Kubes P: Intraluminal crawling of neutrophils to emigration sites: a molecularly distinct process from adhesion in the recruitment cascade. J Exp Med 2006;203:2569–2575.

26 Green CE, Schaff UY, Sarantos MR, Lum AF, Staunton DE, Simon SI: Dynamic shifts in LFA-1 affinity regulate neutrophil rolling, arrest, and transmigration on inflamed endothelium. Blood 2006; 107:2101–2111.
27 Ding Z, Issekutz TB, Downey GP, Waddell TK: L-selectin stimulation enhances functional expression of surface CXCR4 in lymphocytes: implications for cellular activation during adhesion and migration. Blood 2003;101:4245–4252.
28 Cinamon G, Shinder V, Alon R: Shear forces promote lymphocyte migration across vascular endothelium bearing apical chemokines. Nat Immunol 2001; 2:515–522.
29 Phillips R, Ager A: Activation of pertussis toxin-sensitive CXCL12 (SDF-1) receptors mediates transendothelial migration of T lymphocytes across lymph node high endothelial cells. Eur J Immunol 2002;32:837–847.
30 Chen A, Engel P, Tedder TF: Structural requirements regulate endoproteolytic release of the L-selectin (CD62L) adhesion receptor from the cell surface of leukocytes. J Exp Med 1995;182:519–530.
31 Migaki GI, Kahn J, Kishimoto TK: Mutational analysis of the membrane-proximal cleavage site of L-selectin: relaxed sequence specificity surrounding the cleavage site. J Exp Med 1995;182:549–557.
32 Stoddart JH Jr, Jasuja RR, Sikorski MA, von Andrian UH, Mier JW: Protease-resistant L-selectin mutants. Down-modulation by cross-linking but not cellular activation. J Immunol 1996;157:5653–5659.
33 Zhao LC, Edgar JB, Dailey MO: Characterization of the rapid proteolytic shedding of murine L-selectin. Dev Immunol 2001;8:267–277.
34 Kishimoto TK, Jutila MA, Berg EL, Butcher EC: Neutrophil Mac-1 and MEL-14 adhesion proteins inversely regulated by chemotactic factors. Science 1989;245:1238–1241.
35 Chao CC, Jensen R, Dailey MO: Mechanisms of L-selectin regulation by activated T cells. J Immunol 1997;159:1686–1694.
36 Gomez-Gaviro MV, Dominguez-Jimenez C, Carretero JM, Sabando P, Gonzalez-Alvaro I, Sanchez-Madrid F, Diaz-Gonzalez F: Down-regulation of L-selectin expression in neutrophils by nonsteroidal anti-inflammatory drugs: role of intracellular ATP concentration. Blood 2000;96:3592–3600.
37 Strausbaugh HJ, Rosen SD: A potential role for annexin 1 as a physiologic mediator of glucocorticoid-induced L-selectin shedding from myeloid cells. J Immunol 2001;166:6294–6300.
38 Wang J, Marschner S, Finkel TH: CXCR4 engagement is required for HIV-1-induced L-selectin shedding. Blood 2004;103:1218–1221.
39 Le Gall SM, Bobe P, Reiss K, Horiuchi K, Niu XD, Lundell D, Gibb DR, Conrad D, Saftig P, Blobel CP: ADAMs 10 and 17 represent differentially regulated components of a general shedding machinery for membrane proteins such as transforming growth factor-α, L-selectin, and tumor necrosis factor-α. Mol Biol Cell 2009;20:1785–1794.
40 Kahn J, Walcheck B, Migaki GI, Jutila MA, Kishimoto TK: Calmodulin regulates L-selectin adhesion molecule expression and function through a protease-dependent mechanism. Cell 1998;92:809–818.
41 Peschon JJ, Slack JL, Reddy P, Stocking KL, Sunnarborg SW, Lee DC, Russell WE, Castner BJ, Johnson RS, Fitzner JN, Boyce RW, Nelson N, Kozlosky CJ, Wolfson MF, Rauch CT, Cerretti DP, Paxton RJ, March CJ, Black RA: An essential role for ectodomain shedding in mammalian development. Science 1998;282:1281–1284.
42 Gomez-Gaviro MV, Gonzalez-Alvaro I, Dominguez-Jimenez C, Peschon J, Black RA, Sanchez-Madrid F, Diaz-Gonzalez F: Structure-function relationship and role of tumor necrosis factor-α-converting enzyme in the down-regulation of L-selectin by non-steroidal anti-inflammatory drugs. J Biol Chem 2002;277:38212–38221.
43 Nagano O, Murakami D, Hartmann D, De Strooper B, Saftig P, Iwatsubo T, Nakajima M, Shinohara M, Saya H: Cell-matrix interaction via CD44 is independently regulated by different metalloproteinases activated in response to extracellular Ca^{2+} influx and PKC activation. J Cell Biol 2004;165:893–902.
44 Gomez-Gaviro M, Dominguez-Luis M, Canchado J, Calafat J, Janssen H, Lara-Pezzi E, Fourie A, Tugores A, Valenzuela-Fernandez A, Mollinedo F, Sanchez-Madrid F, Diaz-Gonzalez F: Expression and regulation of the metalloproteinase ADAM-8 during human neutrophil pathophysiological activation and its catalytic activity on L-selectin shedding. J Immunol 2007;178:8053–8063.
45 Jasuja RR, Mier JW: Differential effects of hydroxamate inhibitors on PMA and ligand-induced L-selectin down-modulation: role of membrane proximal and cytoplasmic domains. Int J Immunopathol Pharmacol 2000;13:1–12.
46 Li Y, Brazzell J, Herrera A, Walcheck B: ADAM17 deficiency by mature neutrophils has differential effects on L-selectin shedding. Blood 2006;108:2275–2279.
47 Tu L, Poe JC, Kadono T, Venturi GM, Bullard DC, Tedder TF, Steeber DA: A functional role for circulating mouse L-selectin in regulating leukocyte/endothelial cell interactions in vivo. J Immunol 2002; 169:2034–2043.

48 Lee D, Schultz JB, Knauf PA, King MR: Mechanical shedding of L-selectin from the neutrophil surface during rolling on sialyl Lewis × under flow. J Biol Chem 2007;282:4812–4820.
49 Walcheck B, Alexander SR, St Hill CA, Matala E: ADAM-17-independent shedding of L-selectin. J Leukoc Biol 2003;74:389–394.
50 Nicholson MW, Barclay AN, Singer MS, Rosen SD, Van der Merwe PA: Affinity and kinetic analysis of L-selectin (CD62L) binding to glycosylation-dependent cell-adhesion molecule-1. J Biol Chem 1998; 273:763–770.
51 Donnelly SC, Haslett C, Dransfield I, Robertson CE, Carter DC, Ross JA, Grant IS, Tedder TF: Role of selectins in development of adult respiratory distress syndrome. Lancet 1994;344:215–219.
52 Faveeuw C, Preece G, Ager A: Transendothelial migration of lymphocytes across high endothelial venules into lymph nodes is affected by metalloproteinases. Blood 2001;98:688–695.
53 Klinger A, Gebert A, Bieber K, Kalies K, Ager A, Bell EB, Westermann J: Cyclical expression of L-selectin (CD62L) by recirculating T cells. Int Immunol 2009; 21:443–455.
54 Hafezi-Moghadam A, Ley K: Relevance of L-selectin shedding for leukocyte rolling in vivo. J Exp Med 1999;189:939–948.
55 Walcheck B, Kahn J, Fisher JM, Wang BB, Fisk RS, Payan DG, Feehan C, Betageri R, Darlak K, Spatola AF, Kishimoto TK: Neutrophil rolling altered by inhibition of L-selectin shedding in vitro. Nature 1996;380:720–723.
56 Galkina E, Florey O, Zarbock A, Smith BR, Preece G, Lawrence MB, Haskard DO, Ager A: T lymphocyte rolling and recruitment into peripheral lymph nodes is regulated by a saturable density of L-selectin (CD62L). Eur J Immunol 2007;37:1243–1253.
57 Hafezi-Moghadam A, Thomas KL, Prorock AJ, Huo Y, Ley K: L-selectin shedding regulates leukocyte recruitment. J Exp Med 2001;193:863–872.
58 Allport JR, Ding HT, Ager A, Steeber DA, Tedder TF, Luscinskas FW: L-selectin shedding does not regulate human neutrophil attachment, rolling, or transmigration across human vascular endothelium in vitro. J Immunol 1997;158:4365–4372.
59 Venturi GM, Conway RM, Steeber DA, Tedder TF: CD25+CD4+ regulatory T cell migration requires L-selectin expression: L-selectin transcriptional regulation balances constitutive receptor turnover. J Immunol 2007;178:291–300.
60 Tang ML, Steeber DA, Zhang XQ, Tedder TF: Intrinsic differences in L-selectin expression levels affect T and B lymphocyte subset-specific recirculation pathways. J Immunol 1998;160:5113–5121.
61 Gauguet JM, Rosen SD, Marth JD, von Andrian UH: Core 2 branching β1,6-*N*-acetylglucosaminyltransferase and high endothelial cell *N*-acetylglucosamine-6-sulfotransferase exert differential control over B- and T-lymphocyte homing to peripheral lymph nodes. Blood 2004;104:4104–4112.
62 Hickey MJ, Forster M, Mitchell D, Kaur J, De Caigny C, Kubes P: L-selectin facilitates emigration and extravascular locomotion of leukocytes during acute inflammatory responses in vivo. J Immunol 2000;165:7164–7170.
63 Ng-Sikorski J, Linden L, Eierman D, Franzen L, Molony L, Andersson T: Engagement of L-selectin impairs the actin polymerizing capacity of β_2 integrins on neutrophils. J Cell Sci 1996;109:2361–2369.
64 Rosen SD: Ligands for L-selectin: homing, inflammation, and beyond. Annu Rev Immunol 2004;22: 129–156.
65 Savinov AY, Rozanov DV, Golubkov VS, Wong FS, Strongin AY: Inhibition of membrane type-1 matrix metalloproteinase by cancer drugs interferes with the homing of diabetogenic T cells into the pancreas. J Biol Chem 2005;280:27755–27758.
66 Hundhausen C, Misztela D, Berkhout TA, Broadway N, Saftig P, Reiss K, Hartmann D, Fahrenholz F, Postina R, Matthews V, Kallen KJ, Rose-John S, Ludwig A: The disintegrin-like metalloproteinase ADAM10 is involved in constitutive cleavage of CX3CL1 (fractalkine) and regulates CX3CL1-mediated cell-cell adhesion. Blood 2003;102:1186–1195.
67 Hundhausen C, Schulte A, Schulz B, Andrzejewski MG, Schwarz N, von Hundelshausen P, Winter U, Paliga K, Reiss K, Saftig P, Weber C, Ludwig A: Regulated shedding of transmembrane chemokines by the disintegrin and metalloproteinase 10 facilitates detachment of adherent leukocytes. J Immunol 2007;178:8064–8072.
68 Tsou CL, Haskell CA, Charo IF: Tumor necrosis factor-α-converting enzyme mediates the inducible cleavage of fractalkine. J Biol Chem 2001;276:44622–44626.
69 Schulz B, Pruessmeyer J, Maretzky T, Ludwig A, Blobel CP, Saftig P, Reiss K: ADAM10 regulates endothelial permeability and T-cell transmigration by proteolysis of vascular endothelial cadherin. Circ Res 2008;102:1192–1201.
70 Corada M, Mariotti M, Thurston G, Smith K, Kunkel R, Brockhaus M, Lampugnani MG, Martin-Padura I, Stoppacciaro A, Ruco L, McDonald DM, Ward PA, Dejana E: Vascular endothelial-cadherin is an important determinant of microvascular integrity in vivo. Proc Natl Acad Sci USA 1999;96:9815–9820.

71 Huang AJ, Manning JE, Bandak TM, Ratau MC, Hanser KR, Silverstein SC: Endothelial cell cytosolic free calcium regulates neutrophil migration across monolayers of endothelial cells. J Cell Biol 1993; 120:1371–1380.
72 Tsakadze NL, Sithu SD, Sen U, English WR, Murphy G, D'Souza SE: Tumor necrosis factor-α-converting enzyme (TACE/ADAM-17) mediates the ectodomain cleavage of intercellular adhesion molecule-1 (ICAM-1). J Biol Chem 2006;281:3157–3164.
73 Singh RJ, Mason JC, Lidington EA, Edwards DR, Nuttall RK, Khokha R, Knauper V, Murphy G, Gavrilovic J: Cytokine stimulated vascular cell adhesion molecule-1 ectodomain release is regulated by TIMP-3. Cardiovasc Res 2005;67:39–49.
74 Carman CV, Springer TA: A transmigratory cup in leukocyte diapedesis both through individual vascular endothelial cells and between them. J Cell Biol 2004;167:377–388.
75 Barreiro O, Yanez-Mo M, Serrador JM, Montoya MC, Vicente-Manzanares M, Tejedor R, Furthmayr H, Sanchez-Madrid F: Dynamic interaction of VCAM-1 and ICAM-1 with moesin and ezrin in a novel endothelial docking structure for adherent leukocytes. J Cell Biol 2002;157:1233–1245.
76 Deem TL, Cook-Mills JM: Vascular cell adhesion molecule 1 activation of endothelial cell matrix metalloproteinases: role of reactive oxygen species. Blood 2004;104:2385–2393.
77 Romanic AM, Madri JA: The induction of 72-kDa gelatinase in T cells upon adhesion to endothelial cells is VCAM-1 dependent. J Cell Biol 1994;125: 1165–1178.
78 Wang S, Voisin MB, Larbi KY, Dangerfield J, Scheiermann C, Tran M, Maxwell PH, Sorokin L, Nourshargh S: Venular basement membranes contain specific matrix protein low expression regions that act as exit points for emigrating neutrophils. J Exp Med 2006;203:1519–1532.
79 Murphy G, Murthy A, Khokha R: Clipping, shedding and RIPing keep immunity on cue. Trends Immunol 2008;29:75–82.
80 Kaldjian EP, Gretz JE, Anderson AO, Shi Y, Shaw S: Spatial and molecular organization of lymph node T cell cortex: a labyrinthine cavity bounded by an epithelium-like monolayer of fibroblastic reticular cells anchored to basement membrane-like extracellular matrix. Int Immunol 2001;13:1243–1253.
81 Sabeh F, Shimizu-Hirota R, Weiss SJ: Protease-dependent versus -independent cancer cell invasion programs: three-dimensional amoeboid movement revisited. J Cell Biol 2009;185:11–19.
82 Wolf K, Muller R, Borgmann S, Brocker EB, Friedl P: Amoeboid shape change and contact guidance: T-lymphocyte crawling through fibrillar collagen is independent of matrix remodeling by MMPs and other proteases. Blood 2003;102:3262–3269.
83 Gupta GP, Nguyen DX, Chiang AC, Bos PD, Kim JY, Nadal C, Gomis RR, Manova-Todorova K, Massague J: Mediators of vascular remodelling co-opted for sequential steps in lung metastasis. Nature 2007;446: 765–770.
84 Biancone L, Araki M, Araki K, Vassalli P, Stamenkovic I: Redirection of tumor metastasis by expression of E-selectin in vivo. J Exp Med 1996;183: 581–587.
85 Bos PD, Zhang XH, Nadal C, Shu W, Gomis RR, Nguyen DX, Minn AJ, van de Vijver MJ, Gerald WL, Foekens JA, Massague J: Genes that mediate breast cancer metastasis to the brain. Nature 2009;459: 1005–1009.
86 Borsig L, Wong R, Hynes RO, Varki NM, Varki A: Synergistic effects of L- and P-selectin in facilitating tumor metastasis can involve non-mucin ligands and implicate leukocytes as enhancers of metastasis. Proc Natl Acad Sci USA 2002;99:2193–2198.
87 Laubli H, Stevenson JL, Varki A, Varki NM, Borsig L: L-selectin facilitation of metastasis involves temporal induction of Fut7-dependent ligands at sites of tumor cell arrest. Cancer Res 2006;66:1536–1542.
88 Al-Mehdi AB, Tozawa K, Fisher AB, Shientag L, Lee A, Muschel RJ: Intravascular origin of metastasis from the proliferation of endothelium-attached tumor cells: a new model for metastasis. Nat Med 2000;6:100–102.
89 Wylie S, MacDonald IC, Varghese HJ, Schmidt EE, Morris VL, Groom AC, Chambers AF: The matrix metalloproteinase inhibitor batimastat inhibits angiognesis in liver metastases of B16F1 melanoma cells. Clin Exp Metastasis 1999;17:111–117.
90 Najy AJ, Day KC, Day ML: The ectodomain shedding of E-cadherin by ADAM15 supports ErbB receptor activation. J Biol Chem 2008;283:18393–18401.
91 Najy AJ, Day KC, Day ML: ADAM15 supports prostate cancer metastasis by modulating tumor cell-endothelial cell interaction. Cancer Res 2008;68: 1092–1099.
92 Hidalgo A, Peired AJ, Wild MK, Vestweber D, Frenette PS: Complete identification of E-selectin ligands on neutrophils reveals distinct functions of PSGL-1, ESL-1, and CD44. Immunity 2007;26:477–489.
93 Janes PW, Saha N, Barton WA, Kolev MV, Wimmer-Kleikamp SH, Nievergall E, Blobel CP, Himanen JP, Lackmann M, Nikolov DB: Adam meets Eph: an ADAM substrate recognition module acts as a molecular switch for ephrin cleavage in trans. Cell 2005;123:291–304.

94 Soond SM, Everson B, Riches DW, Murphy G: ERK-mediated phosphorylation of Thr735 in TNF-α-converting enzyme and its potential role in TACE protein trafficking. J Cell Sci 2005;118:2371–2380.

95 Sinclair LV, Finlay D, Feijoo C, Cornish GH, Gray A, Ager A, Okkenhaug K, Hagenbeek TJ, Spits H, Cantrell DA: Phosphatidylinositol-3-OH kinase and nutrient-sensing mTOR pathways control T lymphocyte trafficking. Nat Immunol 2008;9:513–521.

96 Horiuchi K, Le Gall S, Schulte M, Yamaguchi T, Reiss K, Murphy G, Toyama Y, Hartmann D, Saftig P, Blobel CP: Substrate selectivity of epidermal growth factor-receptor ligand sheddases and their regulation by phorbol esters and calcium influx. Mol Biol Cell 2007;18:176–188.

97 Zhao L, Shey M, Farnsworth M, Dailey MO: Regulation of membrane metalloproteolytic cleavage of L-selectin (CD62L) by the epidermal growth factor domain. J Biol Chem 2001;276:30631–30640.

98 Smalley DM, Ley K: L-selectin: mechanisms and physiological significance of ectodomain cleavage. J Cell Mol Med 2005;9:255–266.

99 Murai T, Miyauchi T, Yanagida T, Sako Y: Epidermal growth factor-regulated activation of Rac GTPase enhances CD44 cleavage by metalloproteinase disintegrin ADAM10. Biochem J 2006;395:65–71.

100 Poghosyan Z, Robbins SM, Houslay MD, Webster A, Murphy G, Edwards DR: Phosphorylation-dependent interactions between ADAM15 cytoplasmic domain and Src family protein-tyrosine kinases. J Biol Chem 2002;277:4999–5007.

101 Zhong JL, Poghosyan Z, Pennington CJ, Scott X, Handsley MM, Warn A, Gavrilovic J, Honert K, Kruger A, Span PN Sweep FC, Edwards DR: Distinct functions of natural ADAM-15 cytoplasmic domain variants in human mammary carcinoma. Mol Cancer Res 2008;6:383–394.

102 Maretzky T, Le Gall SM, Worpenberg-Pietruk S, Eder J, Overall CM, Huang XY, Poghosyan Z, Edwards DR, Blobel CP: Src stimulates fibroblast growth factor receptor-2 shedding by an ADAM15 splice variant linked to breast cancer. Cancer Res 2009;69:4573–4576.

103 Richards H, Longhi MP, Wright K, Gallimore A, Ager A: CD62L (L-selectin) down-regulation does not affect memory T cell distribution but failure to shed compromises anti-viral immunity. J Immunol 2008;180:198–206.

Dr. Ann Ager
Department of Infection, Immunity and Biochemistry
School of Medicine, Cardiff University
Heath Park, Cardiff CF14 4XN (UK)
Tel. +44 2920 687 011, Fax +44 2920 687 018/687 303, E-Mail agera@cardiff.ac.uk

Entschladen F, Zänker KS (eds): Cell Migration: Signalling and Mechanisms.
Transl Res Biomed. Basel, Karger, 2010, vol 2, pp 102–119

Guided Tour of Cell Migration: Signals and Pathways

Janina Ratke · Kerstin Lang

Institute of Immunology, Witten/Herdecke University, Witten, Germany

Abstract

Cell migration is not an intrinsic property of the cell, but a process that is regulated by extracellular signal substances and a multitude of external factors from other tissues and organ systems within the body. Accordingly, several studies demonstrate a strong influence of cytokines/chemokines, neurotransmitters and adipocytokines (soluble factors derived from fat cells) on the migratory behavior of immune cells as well as tumor cells. The migration of leukocytes is a key feature to fight cancer cells, whereas the locomotion of tumor cells is a prerequisite for tumor formation and metastasis. The impact of these substances differs depending on the cell type and the signal transduction pathways used. All these signaling molecules bind to varying receptor types and mediate their effects on cell migration via multiple signaling pathways. To illuminate the interplay between the nervous system, the immune system, adipocytes and tumor cells, we herein summarize in vitro and in vivo experiments with regard to cell migration, and deliver insight into the underlying signal transduction pathways.

Cell migration is a vital process involved in normal human development, wound healing and inflammatory responses. The migratory capacity of immune cells is a mandatory component for host immune surveillance. For example, neutrophil granulocytes are quickly recruited from the circulation to migrate into the surrounding tissues to destroy invading microorganisms [1, 2], lymphocytes continuously patrol through the body to detect pathogenic invaders [3], and dendritic cells migrate from inflamed or injured peripheral tissues to secondary lymph nodes to present foreign antigens to naive T cells [4]. In contrast, a dysregulation of cell movement causes several pathological states such as developmental defects, healing abnormalities and cancer metastasis. Most cancer deaths are due to the development of metastases, hence the most important improvement of morbidity and mortality will result from prevention or elimination of such disseminated diseases [5]. Metastasis is a complex series of steps in which cancer cells leave the original tumor site and migrate to other parts of the body via the bloodstream or the lymphatic system. Thereby, migratory activity is not an intrinsic function of the cells, but a process that is regulated by extracellular signal substances and a multitude of external factors from other tissues and organ systems within the body. The immediate environment of tumor cells, i.e. the proteins of the extracellular matrix (ECM), instigate migration [6], as do some ligands of receptor tyrosine kinases, or ligands to cytokine receptors such as leptin [7]. However, the most potent inducers of migration are those ligands which bind to G-protein-coupled receptors (GPCRs), including neurotransmitters and chemokines. A multitude of studies have repeatedly shown that external signal substances such as neurotransmitters and chemokines significantly stimulate the migration of tumor cells as well as leukocytes [8–12]. Moreover, this ligand-induced locomotion differs in leukocytes and tumor cells with regard to migratory dynamics and the intracellular signaling mechanisms [13].

Induction of Migratory Activity by Extracellular Signal Substances

Mediators of Cell Migration: Cytokines

Cytokines encompass a large and structurally diverse family of secreted or membrane-bound proteins that are produced widely throughout the body by cells of diverse embryological origin, and are best known for their many roles in the development and functioning of both the innate and adaptive immune response. Their main task is the regulation of growth, activation and differentiation of immune cells, whereas current studies prove an impact of cytokines on the cancer pathogenesis, too [14]. Because of the pleiotropy and apparent redundancy of cytokine action, a classification of the diverse cytokines and receptor types is difficult. However, based on their presumed function, cells of secretion or target of action they can for example be classed as lymphokines, chemokines, interleukins and adipocytokines. Accordingly, the corresponding cytokine receptors have been divided into several families based on their structural organization and one generally distinguishes between type I cytokine receptors and type II cytokine receptors. The type I cytokine receptor family includes those for interleukin (IL)-2, IL-3, IL-4, IL-6, IL-11, IL-12, granulocyte macrophage colony-stimulating factor receptor (GM-CSF), oncostatin M receptor or leukemia inhibitory factor. Upon ligand binding the receptor molecules form homodimers or heterodimers and the intracellular receptor domains become associated with a variety of signaling molecules such as the Janus kinase (JAK) as tyrosine kinase, and latent cytoplasmatic transcriptional activators such as the signal transducer and activators of transcription (STATs) [15]. Members of the type II cytokine receptor family include primarily those that bind interferons, and intracellularly transduce their signals via the JAK/STAT pathway, too. In contrast, chemokine receptors are all members of the large family of GPCRs, also called serpentine receptors with seven transmembrane domains. In the following we will focus on cytokines affecting the migratory behavior of immune cells and tumor cells.

Cell migration is frequently induced by chemokines that act through GPCRs, and only a few cytokines, signaling through single-transmembrane domain receptors, have been shown to induce cell migration. For example, GM-CSF acts via a member of the type I cytokine receptor family and induces the chemotaxis and chemokinesis of neutrophil granulocytes [16], whereas it diminishes the chemotactic migration to IL-8 but not *N*-formyl-methionyl-leucyl-phenylalanine (fMLP) [17]. IL-6 is a potent stimulus for the chemotaxis of monocytic cells and their transmigration across an endothelial cell layer [18]. Moreover, IL-6 and IL-11 are mediators of T-cell movement, whereas no locomotion was found after stimulation with leukemia inhibitory factor and oncostatin M [19]. In contrast, oncostatin M has been shown to induce chemotaxis of neutrophils [20], and efficient migration of dendritic cells to regional lymph nodes [21]. Furthermore, substances like lactoferrin, a glycoprotein present in milk and neutrophils, mediate indirectly anti-inflammatory activities in vivo by downregulating the cytokine production, for example in monocytic cells [22], or by being a direct chemoattractant of monocytes [23]. Besides their effects on cells of the immune system, all these cytokines also have an impact on tumor progression. Accordingly, oncostatin-M-treated breast carcinoma cells show an increase in the invasive capacity [24, 25], whereas IL-6, in cooperation with an autocrine epidermal growth factor receptor loop, stimulates the migration of breast carcinoma cells [26].

Adipocytokines are a group of novel and highly active molecules, which are abundantly secreted by adipocytes (fat cells), and act at both the local and systemic level [27]. Since their discovery in the early 1990s, when the first member – leptin – was described, around 20 members of the adipocytokine family have been identified so far [28]. In addition to their responsibility to influence energy homeostasis, new studies have identified their decisive role in regulating both adaptive and innate immunity as well as tumor progression, including cell migration respectively. A key molecule in obesity is leptin, a 16-kDa peptide hormone predominantly produced by white adipose tissue [29]. The main function of leptin in the human body is the regulation of energy expenditure and control of appetite. A characteristic of obese individuals is an increase of serum leptin, which is in proportion to body fat mass, e.g. increased in obese and suggesting a loss of the regulation of food intake by this hormone. Leptin acts via transmembrane receptors (OB-R), which belong to the class I cytokine receptor family. The OB-R has at least six isoforms, termed OB-R(a-f), which are generated primarily by alternative splicing of the ob gene whereas only one of them (OB-Rb) has full signaling capabilities and is able to activate the JAK/STAT pathway, the major pathway used by leptin to exert its effects [30]. Leptin is able to stimulate chemokinesis of eosinophils [31], chemotaxis of neutrophils [32], and the migration and invasion of various cells derived from glioma [33], colon carcinoma [7], as well as prostate cancer [34].

The second best investigated adipocytokine is adiponectin, which is a 30-kDa protein secreted exclusively by white adipocytes. Adiponectin is highly abundant in the circulation and has a broad spectrum of biological activities. In contrast to leptin, levels of adiponectin in obese individuals have shown to be decreased even though it comes from adipose tissue. Adiponectin acts through its two receptors, AdipoR1 and AdipoR2, whereas the underlying signal pathways mediating the effects are still as far as possible unknown. In adiponectin-deficient mice, absence of adiponectin was associated with a 2-fold increase in leukocyte rolling and a 5-fold increase in leukocyte adhesion in the microcirculation [35]. In addition, adiponectin markedly inhibits the phagocytotic activity of macrophages and their production of tumor necrosis factor-α in response to lipopolysaccharide stimulation [36]. It suppresses IL-2-enhanced cytotoxic activity of natural killer (NK) cells without affecting basal NK cell cytotoxicity [37]. Although adiponectin is known to affect the functionality of immune cells, so far practically nothing is known about its effects on the migratory activity of immune cells. Several recent studies found that adiponectin suppresses the cell growth in various breast [38] and prostate cancer cell lines [39]. Furthermore, adiponectin stimulates the motility of chondrosarcoma cells [40]. There are a plethora of other adipocytokines such as resistin and visfatin with a described function in the immune system and in tumor progression, whereas the exact effects are still under investigation.

In summary, adipocytokines do not transduce their signals via one receptor type, but via multiple signaling pathways. However, the pathways have started to be solved, but remain incompletely understood.

Chemokines (chemotactic cytokines) are small molecules of approx. 7–10 kDa that form a large cytokine family composed of about 50 members. Approximately 20 different chemokine receptors have already been identified, and they are all members of the GPCR family. Chemokines are defined either based on the number and spacing of cysteine residues in the ligands (C, CC, CXC and CX3C) [41], and/or their function and pattern of expression (homeostatic and inflammatory chemokines) [3]. Inflammatory chemokines are the vast majority and are specialized for the recruitment of immune cells to inflamed regions, while homeostatic chemokines are present in various microenvironments in lymphoid or non-lymphoid tissues, and support trafficking and positioning of cells belonging to the adaptive immune system [3, 42]. The latter are best known for their effects on motility and directional cell migration. An important example for an inflammatory chemokine is IL-8, which is released by macrophages at sites of inflammation and is a chemoattractant for neutrophil granulocytes [43]. Inflammatory chemokines with a chemoattractive function on activated T lymphocytes are e.g. MIP-1α/β (macrophage inflammatory protein-1α/β), MCP-1 to -4 (monocyte chemotactic protein-1 to -4), and RANTES (regulated upon activation normal T-cell expressed and secreted) [44, 45]. The stromal cell-derived factor (SDF-1) is the most prominent representative for a homeostatic chemokine, and it is the most efficacious chemoattractant for lymphocytes and monocytes [46]. In addition, SDF-1 is a chemoattractant for hematopoietic progenitor and stem cells [47], and it stimulates breast cancer cells to undergo directional migration and to successfully penetrate ECM for invasion [11]. The latter study forms the basis of a review by

Moore [48] which elucidates a locational model of cancer metastases, wherein the metastatic migration of tumor cells to preferential organs is regulated by chemokines. Thus, tumor cells, which harbor specific chemokine receptors, migrate to particular organs, where their respective chemokine ligands are secreted. This view is further supported by various studies demonstrating that human tumor cells frequently produce and release CXC chemokines such as IL-8. Likewise, expression of the appropriate receptors for IL-8 CXCR1 and CXCR2 has been detected on melanoma [49], ovarian carcinoma [50], and bladder carcinoma cells [10]. Human T24 bladder carcinoma cells express both IL-8 receptors and secrete IL-8 which can then act in an autocrine fashion as stimulator for migration and proliferation [10]. The CC chemokine RANTES and its receptors are expressed by breast tumor cells (as measured in biopsy sections), too, and MCF-7 breast carcinoma cells migrate in response to RANTES [51].

In summary, chemokines not only play a key role in host defense mechanisms through their effects on neutrophil activation and leukocyte trafficking, but have a strong impact on the regulation of tumor cell migration and proliferation, too.

Neurotransmitters have been traditionally defined as chemical messengers, which released from a neuron diffuse across a synaptic cleft to bind and stimulate a postsynaptic cell. In contrast to chemokines, neurotransmitters do not form a family of structurally related molecules. However, they can be classified based on their structure in four groups [52]: (1) Biogenic amines, which are modified amino acids: catecholamines, dopamine, serotonin, histamine. (2) Neuropeptides, e.g. natural opiates (endorphin and enkephalins), gut-brain peptides (somatostatin, cholecystokinin, vasoactive intestinal polypeptide) and inflammatory peptides (bradykinin, calcitonin gene-related peptide). (3) Amino acids such as glutamate and γ-aminobutyric acid (GABA). (4) Structurally unrelated molecules, e.g. acetylcholine and anandamide.

The biogenic amines are derivates of amino acids, such as the catecholamines norepinephrine and dopamine are derivates of tyrosine. Catecholamines are also termed stress hormones, since they are released during stress reactions. Catecholamines increase the frequency and intensity of heart muscle contraction, lead to the dilation of airways and have a glycogenolytic effect. Dopamine, a precursor in the synthesis of catecholamines, is synthesized in the brain and in other areas of the central nervous system (CNS) and peripheral nervous system.

Catecholamines have strong, however diverse impacts on the function of various leukocytes, and lymphatic organs are directly innervated by noradrenergic nerve fibers [53]. Depending on the type of immune cell and its activation status, norepinephrine and dopamine do have divergent effects on the migratory activity. Both neurotransmitters increase the spontaneous migratory activity of naive cytotoxic T lymphocytes (CTLs) with dopamine being the strongest inducer, whereas activated CTLs show a reduced migratory activity in the presence of norepinephrine [unpubl. data]. In NK cells, norepinephrine mediates a pro-migratory effect, whereas dopamine does not have any effect [9].

In parallel, norepinephrine and dopamine are the strongest inducers of breast carcinoma cell migration [54]. Moreover, norepinephrine stimulates the locomotory activity of tumor cells of cell lines derived from prostate and colon carcinomas [55, 56], and breast carcinoma cells even show positive chemotaxis towards norepinephrine [54]. In athymic BALB/c nude mice, norepinephrine enhances the development of lumbar lymph node metastases from prostate carcinoma cells injected in the thighs, which is inhibited by specific β-blockers [57]. Based on the fact that noradrenergic nerve fibers are found to innervate the spleen, thymus, bone marrow, and lymph nodes [53], one might even argue that the localization of metastases can be driven by neurotransmitters. For example, metastasis of small cell lung carcinoma can be found primarily in the catecholamine-producing adrenal glands [58], and in the brain. In addition, direct neuropeptidergic innervation of tumors has been observed in human esophageal and cardiac carcinoma [59] and human urinary bladder carcinoma [60]. In this context a new theory exists that tumors initiate their own innervation by the release of neurotrophic factors including the nerve growth factor, the brain-derived growth factor, and the vascular endothelial growth factor. By this process, which is termed neoneurogenesis, the tumor cells get in close contact to the nerve cells, forming a neuro-neoplastic synapse. Through these synapses, neurotransmitters are directly supplied to the tumors which has an impact on tumor growth and metastasis formation [61].

Substance P is another neuropeptide which plays a role in stress reactions, in the regulation of affective behavior as well as in anxiety and depression [62]. Substance P belongs to the neurokinin family, localized in the CNS and peripheral nervous system [63]. This neurotransmitter inhibits the migration of NK cells and activated CTLs [9], whereas naive CTLs show an enhanced migratory activity in response to substance P [unpubl. data]. Substance P was shown to directly influence neutrophil adhesion to and subsequent migration across a subendothelial barrier of fibroblasts and ECM towards lung inflammatory sites [64], and on the other hand neutrophil transendothelial migration is indirectly increased by stimulation of human umbilical vein endothelial cells through an ICAM-1-dependent mechanism [65]. Substance P on one side reduces the invasive potential of PC-3 prostate carcinoma and murine colon carcinoma cells, but also induces the migratory activity of human colon [52] and breast carcinoma cells [55].

The second important group of neurotransmitters with regard to the induction of cell migration consists of those with a function in inflammatory processes. Here, three structurally unrelated neurotransmitters are to be discussed: the biogenic amine histamine, the neuropeptides bradykinin and calcitonin gene-related peptide (CGRP). As part of an immune response to foreign pathogens, histamine is produced by basophils and by mast cells found in nearby connective tissues. Histamine increases the permeability of the capillaries to white blood cells and other proteins, in order to allow them to engage foreign invaders in the affected tissues. It stimulates the locomotion of neutrophil granulocytes [66] and monocytes [67]. Bradykinin and CGRP are the so-called inflammatory neuropeptides. Bradykinin chemotactically recruits neutrophil granulocytes to sites of inflammation [68], but has no effects on the migration of NK cells and T lymphocytes [9]. In contrast, CGRP is a chemoattractant for T lymphocytes [69], and it has a pro-migratory effect on neutrophil granulocytes [70]. Besides their effects on cells of the immune system, these neurotransmitters were also shown to have an impact on tumor progression. Accordingly, histamine is a chemoattractant for melanoma cells [71] and induces the migration of breast carcinoma cells [72], whereas bradykinin and CGRP both enhance the cell motility and invasion of PC-3 prostate cancer cells [73, 74].

Pro-opiomelanocortin is precursor of several neuropeptides, such as adrenocorticotropic hormone, α-melanocyte-stimulating hormone (MSH), and the opioid peptides, which are the endorphins, enkephalins and dynorphins. All these peptides arise from the proteolytic cleavage of pro-opiomelanocortin. In turn, endogenous and exogenous opioids are known to exert direct effects on the immune system and the expression of functional opioid receptors has been reported for several immune cell types. α-MSH has potent anti-inflammatory effects in all animal models of inflammation, e.g. by reducing the fMLP- and IL-8- induced migration of neutrophils [75]. Met-enkephalin, but not β-endorphin, is a strong stimulator for the migration of MDA-MB-468 breast carcinoma cells [54]. α-MSH reduces cell migration and invasion in melanoma cells [76] and reduces uveal melanoma invasion through fibronectin [77].

Another group of neurotransmitters with an impact on cell locomotion are the amino acids such as glutamate and GABA. Glutamate is the predominant excitatory neurotransmitter of the CNS, and is involved in several central neuronal functions such as learning and memory, neuronal development and neuronal degeneration [78, 79]. It mediates its modulatory effects via two types of receptors, the G-protein-coupled metabotropic and the ionotropic glutamate receptors [80]. There are some reports that describe glutamate as key modulator in the immune response of the CNS, but reports on the function of immune cells, especially in peripheral tissue, are rare [81]. However, glutamate was shown to trigger human T-cell function and to increase CXCR4-mediated T-cell chemotactic migration [82]. Likewise, there exist only a few studies demonstrating a potential role of the glutamatergic system in cancer biology. Nevertheless, glioma cells respond to glutamate with an increased migration [83], and glutamate antagonists produce morphological alterations in various tumor cells leading to a decrease in their motility and invasive growth [84].

GABA is synthesized from glutamate by decarboxylation, and the major inhibitory neurotransmitter of the CNS, where it has been shown to play a role in diseases like epilepsy [85]. In the immune system, GABA functions as an inhibitor for the locomotor activity of chemokine-induced migration of CTLs, whereas migratory activity of neutrophil granulocytes was not affected [52], and the cytotoxicity of NK cells seems to be slightly increased [9]. In colon and breast carcinoma cells, GABA inhibits the norepinephrine-induced migratory activity of these cells by the engagement of the G-protein-coupled $GABA_B$ receptor [54, 86]. Signaling via this receptor in lung adenocarcinoma strongly blocks p-extracellular regulated kinase-1 (ERK1)/2 and cell migration [12].

Another neurotransmitter with an inhibitory function on cell migration is anandamide, which belongs to the group of non-related neurotransmitters. Anandamide is an endogenous cannabinoid neurotransmitter, found especially in the brain. It is an arachidonic acid derivate, which binds with varying affinity to both G_i-protein-coupled cannabinoid receptors CB_1-R and CB_2-R [87]. With regard to the immune system, anandamide has an anti-inflammatory function and plays a role in the reduction of chronic pain. For example, anandamide inhibits the fMLP-induced migration of human neutrophils [88], and the chemokine-stimulated migration of T lymphocytes [89]. In parallel, anandamide has an inhibitory function on the migration of human breast cancer cells [90] and colon carcinoma cells [89]. In addition, this cannabinoid was shown to decrease cancer cell invasion by an increased expression of tissue inhibitors of matrix metalloproteinases [91].

Pathways Leading to Cell Migration (Cytokines and Neurotransmitters)

Chemokines and neurotransmitters primarily mediate their effects on the migratory behavior of cells via signaling through GPCRs. GPCRs span the cell membrane and transduce extracellular messages from soluble ligands binding at the cell surface into intracellular second messengers. These messengers initiate signaling cascades that ultimately control myriad cell responses. GPCRs classically transmit their signal via the activation of the intracellularly coupled heterotrimeric G-proteins. With regard to tumor cell migration, ligand binding activates two important pathways (fig. 1). Activation of the G-protein through the receptor causes its dissociation into a GTP-bound α-subunit and a βγ-subunit, whereas each of these parts is signaling in an independent route [92]. The adenylyl cyclase is a key target molecule of the α-subunit. Depending on the type of GPCR, stimulatory (Gs) α-subunits or inhibitory (Gi) α-subunits are activated, resulting in a stimulation or inhibition, respectively, of the adenylyl cyclase, thereby regulating the generation of the second messenger cyclic adenosine monophosphate (cAMP) from ATP [92]. For example, norepinephrine binds to the β_2-adrenoreceptor which is coupled to Gs-proteins and mediate a pro-migratory effect, whereas the cannabinoid receptors and the $GABA_B$ receptor are coupled to Gi-proteins and their ligands, anandamide and GABA, have an inhibitory effect on migration [86, 89]. cAMP in turn mediates its action through the exchange protein directly activated by cAMP (Epac) and the protein kinase A (PKA), which has a multitude of downstream targets involved in the regulation of migration. For example, the PKA is involved in the regulation of the cytosolic calcium concentration by phosphorylating phospholamban, an inhibitory protein of the sarcoplasmatic/endoplasmatic reticulum ATPase [10], the activation of actin filament assembly via

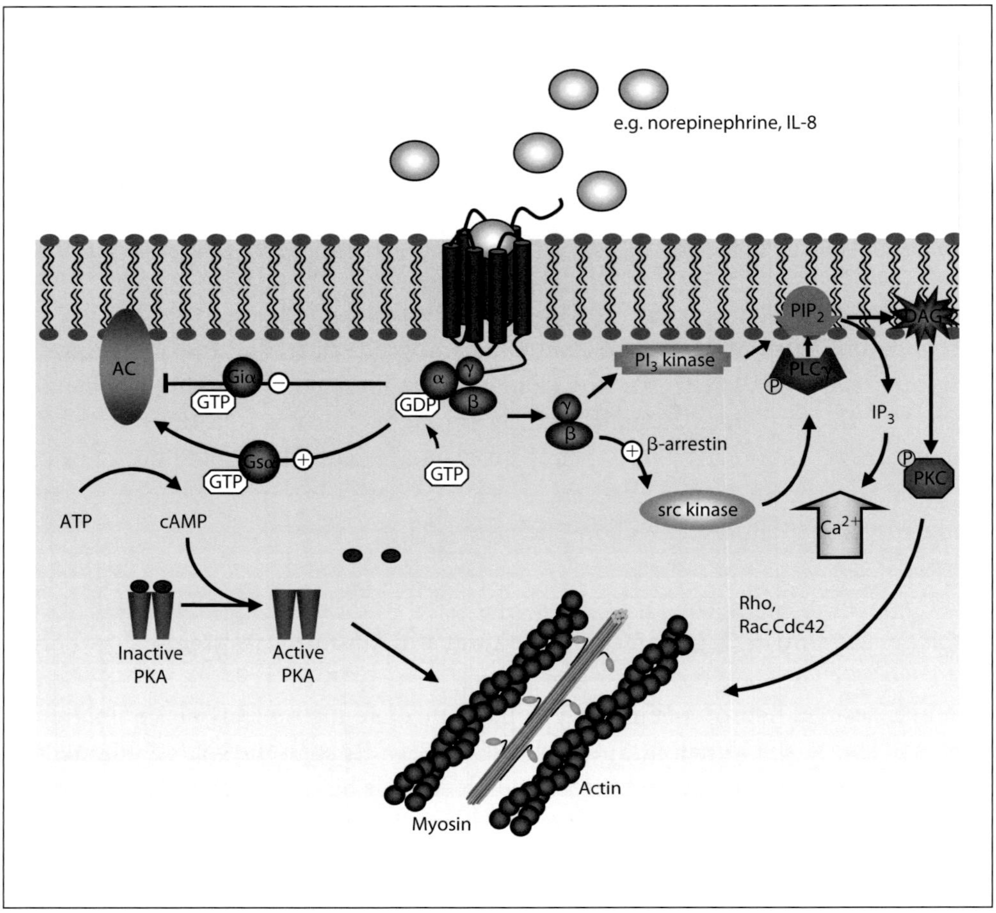

Fig. 1. Simplified scheme of GPCR signaling in cell migration.

action on Ena/VASP proteins and profilin [93, 94], and the activity of myosin, which is discussed in another chapter of this book. Epac is also involved in the control of cell migration via an activation of Rap1 GTPase [95].

The second pathway which is mediated by the βγ-subunit of the G-protein activates GPCR tyrosine kinases, which engage src-protein tyrosine kinases via β-arrestin (fig. 1). This activation leads to the phosphorylation of phospholipase Cγ (PLCγ), which then leads to the transformation of membrane phosphatidyl-inositol-bisphosphate (PIP_2) into inositol triphosphate (IP_3), a second messenger which opens intracellular calcium channels, and diacylglycerol, an activator of classical and novel protein kinase C (PKC) isotypes such as (PKCα) [56]. PKC appears to promote actin polymerization via Rho, Rac and Cdc42, resulting in the formation of membrane ruffles, cell adhesion and actin plaque assemblies [96]. In addition, PKCs are known to regulate proteins that interact with the actin cytoskeleton such as myristoylated, alanine-rich C-kinase substrate and gelsolin, but also by participating in focal adhesion formation [97]. These focal adhesion contacts are large, dynamic multiprotein complexes through which the intracellular cytoskeleton of a cell connects to the ECM. The dynamic assembly and disassembly of focal adhesion plays a central role for cell migration, and the connection between focal adhesions and the ECM generally involves the integrin receptors [93]. Within the cell, the intracellular domain of integrins binds to the actin cytoskeleton via structural proteins like talin, vinculin, and α-actinin [98, 99]. The binding capacities of these proteins are regulated by many other signaling proteins, such as focal adhesion kinase (FAK), PKC, and the Rho family of G-proteins [93, 100, 101]. FAK can be activated by integrin clustering which leads to autophosphorylation at Tyr397, which is a binding site for src family kinases and phosphatidyl-inositol-3-kinase (PI3K) [102]. Interestingly, focal adhesion contacts are an essential structural element in slow-moving cells like fibroblasts and tumor cells, but not in fast-moving cells such as leukocytes. Migrating T lymphocytes do not develop focal adhesions [103], but use more diffuse and highly labile contacts [104]. However, FAK is phosphorylated in these locomoting T cells, suggesting a functional involvement of this kinase in these cells [105]. The PI3K is another key molecule regulating cell migration with a divergent involvement in different cell types. Whereas the migration of CTLs and breast carcinoma cells is impaired by the inhibition of the PI3K, locomotion of neutrophil granulocytes is only slightly affected [13]. Activation of PI3K is facilitated by the βγ-subunit of activated heterotrimeric G-proteins [106], non-receptor protein tyrosine kinases of the src family [107] or FAK [102]. In general, activated PI3Ks phosphorylate PIP_2 to PIP_3 (phosphatidyl-inositol-3,4,5-phosphate), which in turn functions as an adaptor molecule for other signaling molecules such as PKC isoforms [108], and the kinase Akt/protein kinase B. The latter phosphorylate PAKa which is essential for the assembly of myosin II, another important motor protein with a known function for the locomotion of tumor cells and leukocytes as well [109]. Furthermore, girdin is a novel actin-binding Akt substrate that plays an important role in actin organization and Akt-dependent cell motility in fibroblasts and a variety of cancer cell lines [110]. Thus, PI3K mediates its effects via Akt on the two major cellular motor proteins, myosin and actin.

In contrast to GPCRs, members of the cytokine receptor superfamily are cell surface glycoproteins that function as oligomeric complexes consisting of typically two to four receptor chains [111]. Herein we will focus on the signaling mechanisms of the members of the type I and type II cytokine receptor family, because of being more relevant for cell migration. The JAK/signal transducers and activators of transcription (STAT) pathway is the principal signaling mechanism for a wide array of cytokines and growth factors. Mechanistically, JAK/STAT signaling is relatively simple, with only a few principal components. Activation of all known cytokine receptors induces the tyrosine phosphorylation and activation of one or more JAKs associated with the receptor, and JAK activation is required for most if not all receptor functions. In mammals, the JAK family comprises four members: JAK1, JAK2, JAK3 and Tyk2

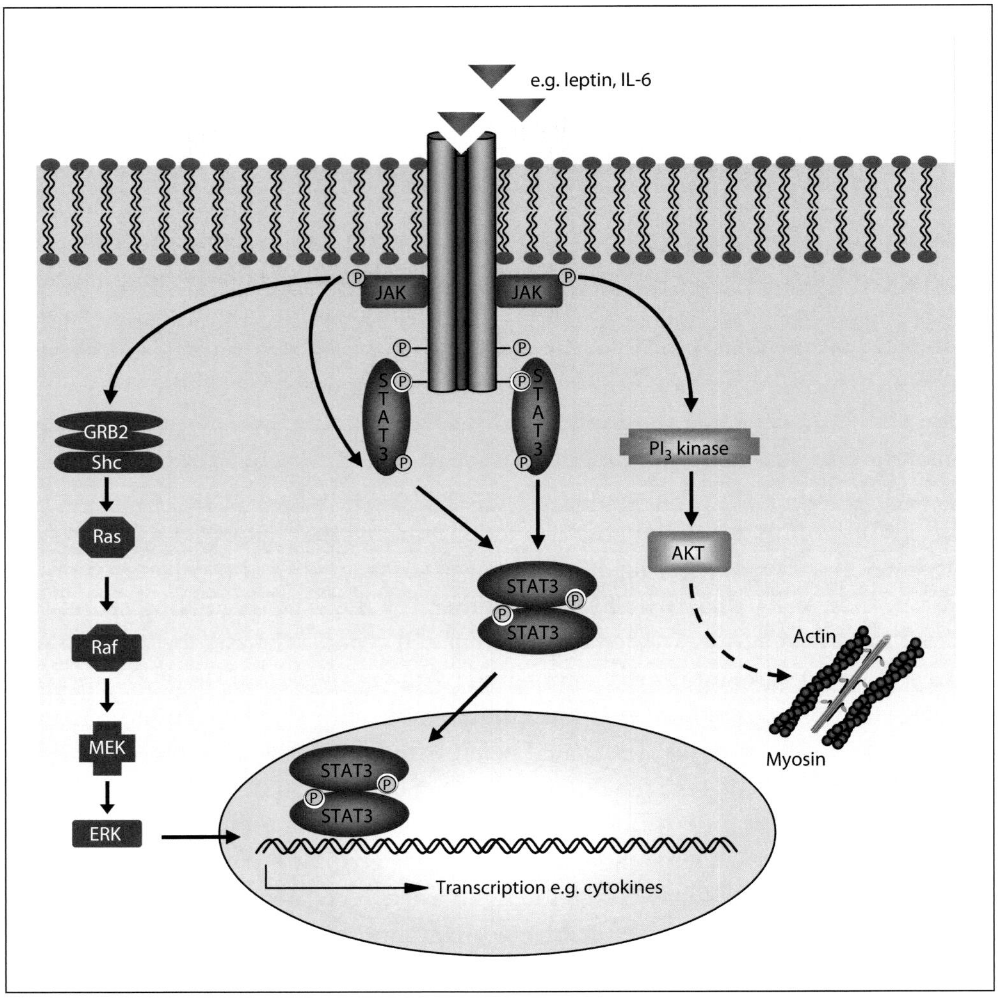

Fig. 2. Simplified scheme of cytokine receptor signaling in cell migration.

[112]. JAK activation occurs upon ligand-mediated receptor multimerization because two JAKs are brought into close proximity, allowing trans-phosphorylation (fig. 2). The activated JAKs subsequently phosphorylate tyrosines on their associated receptors that can serve as docking sites for SH2-containing adaptor proteins from other signaling pathways recruited to the receptor complex such as cytosolic STATs [112]. STATs are latent transcription factors that reside in the cytoplasm until activated. The seven mammalian STATs bear a conserved tyrosine residue near the C-terminus that is phosphorylated by JAKs. Phosphorylation of STATs results in their homo- or heterodimerization and is regulated by SH2 domain interactions [113]. Following phosphorylation, these transcription factors then translocate to the nucleus, activating

Table 1. Ligands with effects on the migration of immune cells and tumor cells

	Effect on tumor cells	Effect on immune cells
Cytokines		
e.g. IL-6, GM-CSF, Oncostatin M	In cooperation with an autocrine EGFR-loop IL-6 stimulates breast carcinoma cell migration	IL-6 mediates chemotaxis of monocytic cells and T-cell movement
	Oncostatin M increases invasive capacity of breast carcinoma [24, 25]	GM-CSF induces chemotaxis and chemokinesis of neutrophils
Adipocytokines		
Leptin	Stimulates migration and invasion of glioma [7] colon carcinoma and prostate cancer cells [34]	Stimulates chemokinesis of eosinophils and chemotaxis of neutrophils [32]
Adiponectin	Induces motility of chondrosarcoma cells [40]	Inhibits phagocytotic activity of macrophages [36] and IL-2 induced cytotoxic activity of NK cells [37]
Chemokines		
IL-8	Stimulates migration of bladder carcinoma cells [10]	Chemoattractant of neutrophil granulocytes
MIP-1α/β, MCP1/4, SDF-1, RANTES	RANTES and SDF-1 induce breast cancer cell migration [11, 51]	SDF is a chemoattractant for lymphocytes, monocytes and stem cells [47]
Neurotransmitters		
Norepinephrine	Stimulates migration of carcinoma cells [55, 56] and lymph node metastasis development	Pro-migratory effect on CTLs and NK cells
Dopamine	Increases breast carcinoma cell locomotion	Enhances migration of CTLs
Histamine	Chemoattractant for melanoma [71] and carcinoma	Stimulates locomotion of monocytes [67] and neutrophils [66]
Substance P	Reduces invasive potential of prostate carcinoma cells, and induces migratory activity of colon [52] and breast carcinoma cells [55]	Inhibits migration of activated CTLs and NK cells
GABA	Inhibitor of norepinephrine-induced colon and breast carcinoma migration	Affects locomotory activity and cytotoxicity of CTLs and NK cells [9, 52]
Anandamide	Decreases cancer cell invasion and migration	Inhibits induced migration of neutrophils and T cells

transcription by binding to the g-activating sequence motif of the promoter regions of various genes [114, 115]. Thus the JAK/STAT pathway represents a mechanism for the rapid transduction of cytokine-induced signals to the nucleus for the activation of transcription. Although the mechanism of JAK/STAT signaling is relatively simple in theory, the biological consequences of pathway activation are complicated by interactions with other signaling pathways [116]. Although these cytokine-signaling pathways are yet to be fully characterized, Ras/MAPK (mitogen-activated protein kinase), Rho and Rac GTPases are the most prominent molecules known to interact with the JAK/STAT-signaling pathways [114]. Most cytokines activate Ras, the p85 subunit of the PI3K, and less often the PLCγ, the last two of which are recruited to the receptor by virtue of their SH2 domain [111]. As we have written previously, all of these molecules play a role in the regulation of cell migration. Crosstalk between the JAK/STAT and Ras/MAPK pathways also involves the activation of transcription proteins, including the transcriptional regulator c-fos [114, 115]. Altogether, the use of the Ras/MAPK, Rho and Rac, and PLC-inositol phosphate cascades by both cytokine-, chemokine- and neurotransmitter-signaling pathways provides potential mechanisms whereby these molecules can cooperatively interact to regulate cell migration.

Conclusion

Signal substances of the nervous system, the immune system and adipose tissue, namely neurotransmitters, cytokines/chemokines and adipocytokines, have a strong impact on the migration of tumor cells and immune cells (table 1). There are multiple mechanisms that integrate the cellular effects of these signaling molecules. The knowledge of the intracellular signal transduction pathways that regulate the migratory activity of tumor cells and leukocytes contributes to the understanding of the complex signaling network, in which metastasis formation and immune response coordination are embedded, and might provide new sources for the specific inhibition of cancer progression towards invasion and metastasis.

Acknowledgement

This work was supported by the Network of Complement Related Disease (Luzern, Switzerland).

References

1 Choi EY, Santoso S, Chavakis T: Mechanisms of neutrophil transendothelial migration. Front Biosci 2009;14:1596–1605.

2 Ley K, Laudanna C, Cybulsky MI, Nourshargh S: Getting to the site of inflammation: the leukocyte adhesion cascade updated. Nat Rev Immunol 2007; 7:678–689.

3 Moser B, Loetscher P: Lymphocyte traffic control by chemokines. Nat Immunol 2001;2:123–128
4 Martin-Fontecha A, Lanzavecchia A, Sallusto F: Dendritic cell migration to peripheral lymph nodes. Handb Exp Pharmacol 2009;188:31–49.
5 Eccles SA, Welch DR: Metastasis: recent discoveries and novel treatment strategies. Lancet 2007;369: 1742–1757.
6 Juliano RL, Haskill S: Signal transduction from the extracellular matrix. J Cell Biol 1993;120:577–585.
7 Jaffe T, Schwartz B: Leptin promotes motility and invasiveness in human colon cancer cells by activating multiple signal-transduction pathways. Int J Cancer 2008;123:2543–2556.
8 Cohen-Hillel E, Mintz R, Meshel T, Garty BZ, Ben-Baruch A: Cell migration to the chemokine CXCL8: paxillin is activated and regulates adhesion and cell motility. Cell Mol Life Sci 2009;66:884–899.
9 Lang K, Drell TL 4th, Niggemann B, Zänker KS, Entschladen F: Neurotransmitters regulate the migration and cytotoxicity in natural killer cells. Immunol Lett 2003;90:165–172.
10 Lang K, Niggemann B, Zanker KS, Entschladen F: Signal processing in migrating T24 human bladder carcinoma cells: role of the autocrine interleukin-8 loop. Int J Cancer 2002;99:673–680.
11 Muller A, Homey B, Soto H, Ge N, Catron D, Buchanan ME, McClanahan T, Murphy E, Yuan W, Wagner SN, Barrera JL, Mohar A, Verastegui E, Zlotnik A: Involvement of chemokine receptors in breast cancer metastasis. Nature 2001;410:50–56.
12 Schuller HM, Al-Wadei HA, Majidi M: γ-Aminobutyric acid, a potential tumor suppressor for small airway-derived lung adenocarcinoma. Carcinogenesis 2008;29:1979–1985.
13 Bastian P, Posch B, Lang K, Niggemann B, Zaenker KS, Hatt H, Entschladen F: Phosphatidylinositol 3-kinase in the G-protein-coupled receptor-induced chemokinesis and chemotaxis of MDA-MB-468 breast carcinoma cells: a comparison with leukocytes. Mol Cancer Res 2006;4:411–421.
14 Dranoff G: Cytokines in cancer pathogenesis and cancer therapy. Nat Rev Cancer 2004;4:11–22.
15 Heim MH: The JAK-STAT pathway: cytokine signalling from the receptor to the nucleus. J Recept Signal Transduct Res 1999;19:75–120.
16 Gomez-Cambronero J: Rapamycin inhibits GM-CSF-induced neutrophil migration. FEBS Lett 2003;550:94–100.
17 Shen L, Smith JM, Shen Z, Hussey SB, Wira CR, Fanger MW: Differential regulation of neutrophil chemotaxis to IL-8 and fMLP by GM-CSF: lack of direct effect of oestradiol. Immunology 2006;117: 205–212.
18 Clahsen T, Schaper F: Interleukin-6 acts in the fashion of a classical chemokine on monocytic cells by inducing integrin activation, cell adhesion, actin polymerization, chemotaxis, and transmigration. J Leukoc Biol 2008;84:1521–1529.
19 Weissenbach M, Clahsen T, Weber C, Spitzer D, Wirth D, Vestweber D, Heinrich PC, Schaper F: Interleukin-6 is a direct mediator of T-cell migration. Eur J Immunol 2004;34:2895–2906.
20 Modur V, Feldhaus MJ, Weyrich AS, Jicha DL, Prescott SM, Zimmerman GA, McIntyre TM: Oncostatin M is a proinflammatory mediator. In vivo effects correlate with endothelial cell expression of inflammatory cytokines and adhesion molecules. J Clin Invest 1997;100:158–168.
21 Sugaya M, Fang L, Cardones AR, Kakinuma T, Jaber SH, Blauvelt A, Hwang ST: Oncostatin M enhances CCL21 expression by microvascular endothelial cells and increases the efficiency of dendritic cell trafficking to lymph nodes. J Immunol 2006;177: 7665–7672.
22 Haversen L, Ohlsson BG, Hahn-Zoric M, Hanson LA, Mattsby-Baltzer I: Lactoferrin down-regulates the LPS-induced cytokine production in monocytic cells via NF-κB. Cell Immunol 2002;220:83–95.
23 De la Rosa G, Yang D, Tewary P, Varadhachary A, Oppenheim JJ: Lactoferrin acts as an alarmin to promote the recruitment and activation of APCs and antigen-specific immune responses. J Immunol 2008;180:6868–6876.
24 Jorcyk CL, Holzer RG, Ryan RE: Oncostatin M induces cell detachment and enhances the metastatic capacity of T-47D human breast carcinoma cells. Cytokine 2006;33:323–336.
25 Queen MM, Ryan RE, Holzer RG, Keller-Peck CR, Jorcyk CL: Breast cancer cells stimulate neutrophils to produce oncostatin M: potential implications for tumor progression. Cancer Res 2005;65:8896–8904.
26 Badache A, Hynes NE: Interleukin-6 inhibits proliferation and, in cooperation with an epidermal growth factor receptor autocrine loop, increases migration of T47D breast cancer cells. Cancer Res 2001;61:383–391.
27 Kershaw EE, Flier JS: Adipose tissue as an endocrine organ. J Clin Endocrinol Metab 2004;89:2548–2556.
28 Housa D, Housova J, Vernerova Z, Haluzik M: Adipocytokines and cancer. Physiol Res 2006;55: 233–244.
29 Harvey J, Ashford ML: Leptin in the CNS: much more than a satiety signal. Neuropharmacology 2003;44:845–854.
30 Cirillo D, Rachiglio AM, la Montagna R, Giordano A, Normanno N: Leptin signaling in breast cancer: an overview. J Cell Biochem 2008;105:956–964.

31 Wong CK, Cheung PF, Lam CW: Leptin-mediated cytokine release and migration of eosinophils: implications for immunopathophysiology of allergic inflammation. Eur J Immunol 2007;37:2337–2348.
32 Montecucco F, Bianchi G, Gnerre P, Bertolotto M, Dallegri F, Ottonello L: Induction of neutrophil chemotaxis by leptin: crucial role for p38 and Src kinases. Ann NY Acad Sci 2006;1069:463–471.
33 Yeh WL, Lu DY, Lee MJ, Fu WM: Leptin induces migration and invasion of glioma cells through MMP-13 production. Glia 2009;57:454–464.
34 Frankenberry KA, Somasundar P, McFadden DW, Vona-Davis LC: Leptin induces cell migration and the expression of growth factors in human prostate cancer cells. Am J Surg 2004;188:560–565.
35 Ouedraogo R, Gong Y, Berzins B, Wu X, Mahadev K, Hough K, Chan L, Goldstein BJ, Scalia R: Adiponectin deficiency increases leukocyte-endothelium interactions via upregulation of endothelial cell adhesion molecules in vivo. J Clin Invest 2007;117:1718–1726.
36 Yokota T, Oritani K, Takahashi I, Ishikawa J, Matsuyama A, Ouchi N, Kihara S, Funahashi T, Tenner AJ, Tomiyama Y, Matsuzawa Y: Adiponectin, a new member of the family of soluble defense collagens, negatively regulates the growth of myelomonocytic progenitors and the functions of macrophages. Blood 2000;96:1723–1732.
37 Kim KY, Kim JK, Han SH, Lim JS, Kim KI, Cho DH, Lee MS, Lee JH, Yoon DY, Yoon SR, Chung JW, Choi I, Kim E, Yang Y: Adiponectin is a negative regulator of NK cell cytotoxicity. J Immunol 2006;176:5958–5964.
38 Grossmann ME, Nkhata KJ, Mizuno NK, Ray A, Cleary MP: Effects of adiponectin on breast cancer cell growth and signaling. Br J Cancer 2008;98:370–379.
39 Bub JD, Miyazaki T, Iwamoto Y: Adiponectin as a growth inhibitor in prostate cancer cells. Biochem Biophys Res Commun 2006;340:1158–1166.
40 Chiu YC, Shieh DC, Tong KM, Chen CP, Huang KC, Chen PC, Fong YC, Hsu HC, Tang CH: Involvement of AdipoR receptor in adiponectin-induced motility and $\alpha_2\beta_1$ integrin upregulation in human chondrosarcoma cells. Carcinogenesis 2009;30:1651–1659.
41 Murphy PM: International Union of Pharmacology. XXX. Update on chemokine receptor nomenclature. Pharmacol Rev 2002;54:227–229.
42 Pals ST, de Gorter DJ, Spaargaren M: Lymphoma dissemination: the other face of lymphocyte homing. Blood 2007;110:3102–3111.
43 Lang K, Hatt H, Niggemann B, Zaenker KS, Entschladen F: A novel function for chemokines: downregulation of neutrophil migration. Scand J Immunol 2003;57:350–361.
44 Luster AD: Chemokines-chemotactic cytokines that mediate inflammation. N Engl J Med 1998;338:436–445.
45 Sallusto F, Lanzavecchia A, Mackay CR: Chemokines and chemokine receptors in T-cell priming and Th1/Th2-mediated responses. Immunol Today 1998;19:568–574.
46 Bleul CC, Fuhlbrigge RC, Casasnovas JM, Aiuti A, Springer TA: A highly efficacious lymphocyte chemoattractant, stromal cell-derived factor-1. J Exp Med 1996;184:1101–1109.
47 Hattori K, Heissig B, Rafii S: The regulation of hematopoietic stem cell and progenitor mobilization by chemokine SDF-1. Leuk Lymphoma 2003;44:575–582.
48 Moore MA: The role of chemoattraction in cancer metastases. Bioessays 2001;23:674–676.
49 Haghnegahdar H, Du J, Wang D, Strieter RM, Burdick MD, Nanney LB, Cardwell N, Luan J, Shattuck-Brandt R, Richmond A: The tumorigenic and angiogenic effects of MGSA/GRO proteins in melanoma. J Leukoc Biol 2000;67:53–62.
50 Venkatakrishnan G, Salgia R, Groopman JE: Chemokine receptors CXCR-1/2 activate mitogen-activated protein kinase via the epidermal growth factor receptor in ovarian cancer cells. J Biol Chem 2000;275:6868–6875.
51 Prest SJ, Rees RC, Murdoch C, Marshall JF, Cooper PA, Bibby M, Li G, Ali SA: Chemokines induce the cellular migration of MCF-7 human breast carcinoma cells: subpopulations of tumour cells display positive and negative chemotaxis and differential in vivo growth potentials. Clin Exp Metastasis 1999;17:389–396.
52 Entschladen F, Lang K, Drell TL 4th, Joseph J, Zaenker KS: Neurotransmitters are regulators for the migration of tumor cells and leukocytes. Cancer Immunol Immunother 2002;51:467–482.
53 Felten DL: Direct innervation of lymphoid organs: substrate for neurotransmitter signaling of cells of the immune system. Neuropsychobiology 1993;28:110–112.
54 Drell TL 4th, Joseph J, Lang K, Niggemann B, Zaenker KS, Entschladen F: Effects of neurotransmitters on the chemokinesis and chemotaxis of MDA-MB-468 human breast carcinoma cells. Breast Cancer Res Treat 2003;80:63–70.

55 Lang K, Drell TL 4th, Lindecke A, Niggemann B, Kaltschmidt C, Zaenker KS, Entschladen F: Induction of a metastatogenic tumor cell type by neurotransmitters and its pharmacological inhibition by established drugs. Int J Cancer 2004;112:231–238.

56 Masur K, Niggemann B, Zanker KS, Entschladen F: Norepinephrine-induced migration of SW480 colon carcinoma cells is inhibited by β-blockers. Cancer Res 2001;61:2866–2869.

57 Palm D, Lang K, Niggemann B, Drell TL 4th, Masur K, Zaenker KS, Entschladen F: The norepinephrine-driven metastasis development of PC-3 human prostate cancer cells in BALB/c nude mice is inhibited by β-blockers. Int J Cancer 2006;118:2744–2749.

58 Miyaji N, Miki T, Itoh Y, Shimada J, Takeshita T, Churei H, Nakajo M: Radiotherapy for adrenal gland metastasis from lung cancer: report of three cases. Radiat Med 1999;17:71–75.

59 Lu SH, Zhou Y, Que HP, Liu SJ: Peptidergic innervation of human esophageal and cardiac carcinoma. World J Gastroenterol 2003;9:399–403.

60 Seifert P, Benedic M, Effert P: Nerve fibers in tumors of the human urinary bladder. Virchows Arch 2002; 440:291–297.

61 Entschladen F, Palm D, Niggemann B, Zaenker KS: The cancer's nervous tooth: considering the neuronal crosstalk within tumors. Semin Cancer Biol 2008;18:171–175.

62 Mantyh PW: Neurobiology of substance P and the NK1 receptor. J Clin Psychiatry 2002;63 Suppl 11:6–10.

63 Weihe E, Nohr D, Michel S, Muller S, Zentel HJ, Fink T, Krekel J: Molecular anatomy of the neuroimmune connection. Int J Neurosci 1991;59:1–23.

64 Kahler CM, Pischel A, Kaufmann G, Wiedermann CJ: Influence of neuropeptides on neutrophil adhesion and transmigration through a lung fibroblast barrier in vitro. Exp Lung Res 2001;27:25–46.

65 Nakagawa N, Sano H, Iwamoto I: Substance P induces the expression of intercellular adhesion molecule-1 on vascular endothelial cells and enhances neutrophil transendothelial migration. Peptides 1995;16:721–725.

66 Akdis CA, Blaser K: Histamine in the immune regulation of allergic inflammation. J Allergy Clin Immunol 2003;112:15–22.

67 Singhal PC, Sankaran RT, Nahar N, Shah N, Patel P: Vasoactive agents modulate migration of monocytes across glomerular endothelial cells. J Investig Med 2000;48:110–117.

68 Ehrenfeld P, Millan C, Matus CE, Figueroa JE, Burgos RA, Nualart F, Bhoola KD, Figueroa CD: Activation of kinin B1 receptors induces chemotaxis of human neutrophils. J Leukoc Biol 2006;80:117–124.

69 Foster CA, Mandak B, Kromer E, Rot A: Calcitonin gene-related peptide is chemotactic for human T lymphocytes. Ann NY Acad Sci 1992;657:397–404.

70 Ahluwalia A, Perrett M: Calcitonin gene-related peptides modulate the acute inflammatory response induced by interleukin-1 in the mouse. Eur J Pharmacol 1994;264:407–415.

71 Tilly BC, Tertoolen LG, Remorie R, Ladoux A, Verlaan I, de Laat SW, Moolenaar WH: Histamine as a growth factor and chemoattractant for human carcinoma and melanoma cells: action through Ca^{2+}-mobilizing H1 receptors. J Cell Biol 1990;110: 1211–1215.

72 Medina V, Croci M, Crescenti E, Mohamad N, Sanchez-Jimenez F, Massari N, Nunez M, Cricco G, Martin G, Bergoc R, Rivera E: The role of histamine in human mammary carcinogenesis: H_3 and H_4 receptors as potential therapeutic targets for breast cancer treatment. Cancer Biol Ther 2008;7:28–35.

73 Nagakawa O, Ogasawara M, Murata J, Fuse H, Saiki I: Effect of prostatic neuropeptides on migration of prostate cancer cell lines. Int J Urol 2001;8:65–70.

74 Taub JS, Guo R, Leeb-Lundberg LM, Madden JF, Daaka Y: Bradykinin receptor subtype 1 expression and function in prostate cancer. Cancer Res 2003; 63:2037–2041.

75 Catania A, Rajora N, Capsoni F, Minonzio F, Star RA, Lipton JM: The neuropeptide α-MSH has specific receptors on neutrophils and reduces chemotaxis in vitro. Peptides 1996;17:675–679.

76 Eves PC, MacNeil S, Haycock JW: α-Melanocyte stimulating hormone, inflammation and human melanoma. Peptides 2006;27:444–452.

77 Canton I, Eves PC, Szabo M, Vidal-Vanaclocha F, Sisley K, Rennie IG, Haycock JW, MacNeil S: Tumor necrosis factor-α increases and α-melanocyte-stimulating hormone reduces uveal melanoma invasion through fibronectin. J Invest Dermatol 2003; 121:557–563.

78 Dhami GK, Ferguson SS: Regulation of metabotropic glutamate receptor signaling, desensitization and endocytosis. Pharmacol Ther 2006;111:260–271.

79 Riedel G, Platt B, Micheau J: Glutamate receptor function in learning and memory. Behav Brain Res 2003;140:1–47.

80 Lee HJ, Wall B, Chen S: G-protein-coupled receptors and melanoma. Pigment Cell Melanoma Res 2008;21:415–428.

81 Pacheco R, Gallart T, Lluis C, Franco R: Role of glutamate on T-cell-mediated immunity. J Neuroimmunol 2007;185:9–19.

82 Ganor Y, Besser M, Ben-Zakay N, Unger T, Levite M: Human T cells express a functional ionotropic glutamate receptor GluR3, and glutamate by itself triggers integrin-mediated adhesion to laminin and fibronectin and chemotactic migration. J Immunol 2003;170:4362–4372.

83 Sontheimer H: A role for glutamate in growth and invasion of primary brain tumors. J Neurochem 2008;105:287–295.
84 Rzeski W, Ikonomidou C, Turski L: Glutamate antagonists limit tumor growth. Biochem Pharmacol 2002;64:1195–1200.
85 Bianchi MT, Song L, Zhang H, Macdonald RL: Two different mechanisms of disinhibition produced by $GABA_A$ receptor mutations linked to epilepsy in humans. J Neurosci 2002;22:5321–5327.
86 Joseph J, Niggemann B, Zaenker KS, Entschladen F: The neurotransmitter γ-aminobutyric acid is an inhibitory regulator for the migration of SW480 colon carcinoma cells. Cancer Res 2002;62:6467–6469.
87 Slipetz DM, O'Neill GP, Favreau L, Dufresne C, Gallant M, Gareau Y, Guay D, Labelle M, Metters KM: Activation of the human peripheral cannabinoid receptor results in inhibition of adenylyl cyclase. Mol Pharmacol 1995;48:352–361.
88 McHugh D, Tanner C, Mechoulam R, Pertwee RG, Ross RA: Inhibition of human neutrophil chemotaxis by endogenous cannabinoids and phytocannabinoids: evidence for a site distinct from CB1 and CB2. Mol Pharmacol 2008;73:441–450.
89 Joseph J, Niggemann B, Zaenker KS, Entschladen F: Anandamide is an endogenous inhibitor for the migration of tumor cells and T lymphocytes. Cancer Immunol Immunother 2004;53:723–728.
90 Laezza C, Pisanti S, Malfitano AM, Bifulco M: The anandamide analog, Met-F-AEA, controls human breast cancer cell migration via the RHOA/RHO kinase signaling pathway. Endocr Relat Cancer 2008;15:965–974.
91 Ramer R, Hinz B: Inhibition of cancer cell invasion by cannabinoids via increased expression of tissue inhibitor of matrix metalloproteinases-1. J Natl Cancer Inst 2008;100:59–69.
92 Neer EJ: Heterotrimeric G proteins: organizers of transmembrane signals. Cell 1995;80:249–257.
93 Jockusch BM, Bubeck P, Giehl K, Kroemker M, Moschner J, Rothkegel M, Rudiger M, Schluter K, Stanke G, Winkler J: The molecular architecture of focal adhesions. Annu Rev Cell Dev Biol 1995;11:379–416.
94 Kwiatkowski AV, Gertler FB, Loureiro JJ: Function and regulation of Ena/VASP proteins. Trends Cell Biol 2003;13:386–392.
95 Yarwood SJ: Microtubule-associated proteins regulate cAMP signalling through exchange protein directly activated by cAMP. Biochem Soc Trans 2005;33:1327–1329.
96 Bokoch GM: Chemoattractant signaling and leukocyte activation. Blood 1995;86:1649–1660.
97 Miranti CK, Ohno S, Brugge JS: Protein kinase C regulates integrin-induced activation of the extracellular regulated kinase pathway upstream of Shc. J Biol Chem 1999;274:10571–10581.
98 Clark EA, Brugge JS: Integrins and signal transduction pathways: the road taken. Science 1995;268:233–239.
99 Stossel TP: The E. Donnall Thomas Lecture, 1993. The machinery of blood cell movements. Blood 1994;84:367–379.
100 Burridge K, Petch LA, Romer LH: Signals from focal adhesions. Curr Biol 1992;2:537–539.
101 Cox EA, Sastry SK, Huttenlocher A: Integrin-mediated adhesion regulates cell polarity and membrane protrusion through the Rho family of GTPases. Mol Biol Cell 2001;12:265–277.
102 Chen HC, Appeddu PA, Isoda H, Guan JL: Phosphorylation of tyrosine 397 in focal adhesion kinase is required for binding phosphatidylinositol 3-kinase. J Biol Chem 1996;271:26329–26334.
103 Friedl P, Brocker EB, Zanker KS: Integrins, cell matrix interactions and cell migration strategies: fundamental differences in leukocytes and tumor cells. Cell Adhes Commun 1998;6:225–236.
104 Lee J, Ishihara A, Jacobson K: How do cells move along surfaces? Trends Cell Biol 1993;3:366–370.
105 Entschladen F, Niggemann B, Zanker KS, Friedl P: Differential requirement of protein tyrosine kinases and protein kinase C in the regulation of T cell locomotion in three-dimensional collagen matrices. J Immunol 1997;159:3203–3210.
106 Brock C, Schaefer M, Reusch HP, Czupalla C, Michalke M, Spicher K, Schultz G, Nurnberg B: Roles of G βγ in membrane recruitment and activation of p110 γ/p101 phosphoinositide 3-kinase γ. J Cell Biol 2003;160:89–99.
107 Stephens L, Eguinoa A, Corey S, Jackson T, Hawkins PT: Receptor stimulated accumulation of phosphatidylinositol (3,4,5)-trisphosphate by G-protein-mediated pathways in human myeloid derived cells. EMBO J 1993;12:2265–2273.
108 Sweeney G: Leptin signalling. Cell Signal 2002;14:655–663.
109 Bastian P, Lang K, Niggemann B, Zaenker KS, Entschladen F: Myosin regulation in the migration of tumor cells and leukocytes within a three-dimensional collagen matrix. Cell Mol Life Sci 2005;62:65–76.
110 Jiang P, Enomoto A, Jijiwa M, Kato T, Hasegawa T, Ishida M, Sato T, Asai N, Murakumo Y, Takahashi M: An actin-binding protein Girdin regulates the motility of breast cancer cells. Cancer Res 2008;68:1310–1318.
111 Ihle JN: Cytokine receptor signalling. Nature 1995;377:591–594.

112 Seidel HM, Lamb P, Rosen J: Pharmaceutical intervention in the JAK/STAT signaling pathway. Oncogene 2000;19:2645–2656.

113 Simson L, Foster PS: Chemokine and cytokine cooperativity: eosinophil migration in the asthmatic response. Immunol Cell Biol 2000;78:415–422.

114 Adachi T, Alam R: The mechanism of IL-5 signal transduction. Am J Physiol 1998;275:C623–633.

115 Van der Bruggen T, Caldenhoven E, Kanters D, Coffer P, Raaijmakers JA, Lammers JW, Koenderman L: Interleukin-5 signaling in human eosinophils involves JAK2 tyrosine kinase and Stat1 α. Blood 1995;85:1442–1448.

116 Heinrich PC, Behrmann I, Haan S, Hermanns HM, Muller-Newen G, Schaper F: Principles of interleukin-6-type cytokine signalling and its regulation. Biochem J 2003;374:1–20.

PD Dr. Kerstin Lang
Institute of Immunology, Witten/Herdecke University
Stockumer Strasse 10, DE–58448 Witten (Germany)
Tel. +49 2302 926 183, Fax +49 2302 926 158, E-Mail lang@uni-wh.de

Entschladen F, Zänker KS (eds): Cell Migration: Signalling and Mechanisms.
Transl Res Biomed. Basel, Karger, 2010, vol 2, pp 120–135

Regulation of the E-Cadherin Adhesion Complex in Tumor Cell Migration and Invasion

Andre Menke · Klaudia Giehl

Department of Internal Medicine I, University of Ulm, Ulm, Germany

Abstract

The complex molecular mechanisms leading to tumor progression and acquisition of a metastatic phenotype are only partially understood. In this review we focus on mechanisms involved in the inhibition of intercellular adhesion, especially the regulation of E-cadherin-mediated adherens junctions during epithelial to mesenchymal transition in early invasive processes. Loss of E-cadherin or perturbation of the E-cadherin complex assembly is a key event in epithelial-mesenchymal transition and is directed by a huge number of mechanisms which differ greatly with regard to cell types and tissues. The reduction in intercellular adhesion interferes with tissue integrity and allows cancer cells to disseminate from the primary tumor thereby initiating cancer metastasis.

Cancer arises from a stepwise accumulation of genetic alterations that drive the progressive transformation of normal human cells into highly proliferative and malignant derivatives. This multi-step process results in essential alterations in cell physiology. These changes determine malignant cell growth characterized by: self-sufficiency in growth signals, insensitivity to growth-inhibitory signals, evasion from apoptosis, unlimited replicative potential, aberrant angiogenesis and further on dissemination from the primary tumor, invasion into the surrounding tissue and finally formation of metastases [1]. To characterize this metastatic process, a cascade has been defined which embraces the following steps: (1) tumor angiogenesis; (2) dissemination of tumor cells from the primary tumor mass; (3) migration and invasion through the basement membrane and extracellular matrix; (4) intravasation into blood or lymphatic vessels; (5) extravasation and invasion to target organ(s); (6) development of a secondary tumor/metastasis [2]. The acquisition of a motile behavior early in metastasis depends on the epithelial-mesenchymal transition (EMT), a process especially known in embryonic development, whereby epithelial cells switch to a mesenchymal progenitor-cell phenotype, facilitate detachment and reorganize the epithelial cell sheets during tumor invasion and metastasis [3]. EMT, which is characterized by inactivation of intercellular junctions, particularly E-cadherin-mediated adherens junctions [4, 5], enables tumor cells to detach from the primary tumor and to invade into the surrounding tissues [6]. Thus, induction of EMT represents an essential step during progression from solid, localized tumors to invasive carcinoma [7]. Studies on EMT resulted in the definition of three major changes in the cellular phenotype [summarized in 7]: (1) morphological conversion of the epithelial cells to spindle-shaped mesenchymal cells with migratory protrusions; (2) loss of epithelial cell-cell junction proteins and epithelial intermediate filaments and acquisition of mesenchymal marker proteins such as vimentin, smooth muscle actin and fibronectin, and (3) conversion to motile cells that can invade through the extracellular matrix (ECM). It has to be emphasized that EMT is reversible at many stages thereby enabling cells to redifferentiate when the inducing trigger has vanished. Moreover, whether and to which extent tumor cells undergo this 'classical' EMT in invasion and metastasis is controversially discussed and a comprehensible molecular definition of the EMT program is still elusive [7].

E-Cadherin-Mediated Adherens Junctions

Acquisition of a motile behavior early in metastasis comprises the inactivation of epithelial cell-cell contacts particularly the E-cadherin-mediated adherens junctions, which are a hallmark of epithelial tissues [4, 5]. But, loss of E-cadherin is not uniformly found in carcinoma: whereas a complete deficiency of E-cadherin is found in some carcinomas, others are characterized by loss of the membranous localization of E-cadherin only in the dedifferentiated areas of the tumor particularly near the invasive front. This heterogeneous localization of E-cadherin has been documented for colon and squamous cell carcinoma and less impressive for pancreatic tumors [6, 8, 9].

E-cadherin is a calcium-dependent transmembrane glycoprotein encoded by the *CDH1* gene in humans and expressed in all mammalian epithelia [10, 11]. Figure 1 exemplifies the typical E-cadherin staining in MCF-7 epithelial breast carcinoma cells. E-cadherin belongs to a large superfamily of transmembrane proteins characterized by a varying number of extracellular cadherin repeats and a conserved intracellular domain containing several protein-binding sites [12]. E-cadherin is a key molecule in developing and maintaining cell polarity, in perpetuating mechanical strength of epithelia and in controlling cell survival as well as cell proliferation [13]. Classical cadherins, such as E-cadherin, bind intracellular directly several cytoplasmic proteins including p120ctn, β-catenin and plakoglobin. These proteins mediate the association of E-cadherin with the actin cytoskeleton via binding to α-catenin [14]. The interaction of the E-cadherin adhesion complex with the actin cytoskeleton

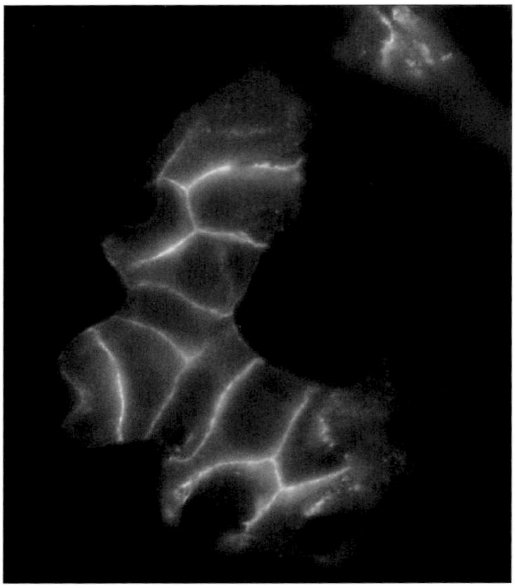

Fig. 1. Immunofluorescence localization of the cell-cell adhesion proteins E-cadherin in cultured MCF-7 breast carcinoma cells. Magnification: 700× (objective 63×).

is induced by a sequential activation of Rac and Rho GTPases and subsequent actin filament polymerization as well as by binding to further, not so well characterized actin-binding proteins such as α-actinin or Eplin [15]. Only E-cadherin which is linked via a catenin-complex to the actin cytoskeleton contributes to strong cellular adhesion [16]. Thus, not all E-cadherin molecules present at the cell membrane are involved in the mediation of mechanical cell-cell adhesion. Perturbation and loss of E-cadherin-mediated adherens junctions as well as reduction of E-cadherin concentration strongly correlates with epithelial dedifferentiation and phenotypic alterations of epithelial cells [17, 18]. Consequently, E-cadherin is considered as a tumor suppressor molecule, because its transcription is downregulated or even completely repressed in various carcinoma [19]. Moreover, re-expression of E-cadherin in carcinoma cells in vitro and in animal models is sufficient to reduce the malignancy of the tumor cells [5, 6]. The assembly and maintenance of adherens junctions is under tight transcriptional and posttranscriptional control. Different mechanisms involved in the repression of E-cadherin-mediated cell-cell adhesion resulting in enhanced migration and malignancy of tumor cells will be outlined below.

Regulation of E-Cadherin Gene Expression

In some human carcinoma, the loss of E-cadherin is due to transcriptional silencing of the E-cadherin promoter by hypermethylation of CpG islands or by histone H3 deacetylation [10]. DNA methylation often occurs close to regulatory promoter regions in CpG islands of the E-cadherin promoter and has been correlated with reduced expression of E-cadherin in various types of cancer [20]. For hepatocellular and breast carcinoma, methylation of the CpG islands in the E-cadherin promoter increases in tumor progression and might be influenced by the tumor microenvironment [21]. A second mechanism in silencing E-cadherin expression are mutations found in E-cadherin itself resulting in premature termination of translation or in deletion of central fragments of the E-cadherin protein. Inactivating mutations were first discovered in gastric cancer [22] but meanwhile described in several other human tumors as well [reviewed in 10]. Moreover, several studies in recent years have demonstrated a loss of heterozygosity of chromosome 16q21-22, the locus of the human E-cadherin gene [reviewed in 10].

Besides epigenetic events, transcriptional repression of E-cadherin expression seems to be a common event in malignant transformation. Different transcription factors have been discussed which bind to the E-cadherin promoter, especially to the E-boxes, which are DNA sequences containing the core sequence CANNTG, resulting in retarded promoter activity. These transcriptional repressors include the basic helix-loop-helix factors E12/E47 and Twist, the two-handed zinc finger homeodomain proteins of the δEF1 family (δEF1/ZEB1 and SIP1/ZEB2), the zinc finger proteins of the Snail family with Snail1 (SNAI1), Slug (SNAI2) and Snail3 (SNAI3) as well as the Lef/TCF family member Lef-1 [11, 23, 24]. Although all of these transcription factors have been shown to be effective in repressing the E-cadherin promoter under certain conditions, a molecular explanation why individual factors are able to repress E-cadherin promoter activity in one epithelial cell line and failed in others is currently missing. But, this knowledge would be of special interest hence most factors bind to or near the E-boxes of the *CDH1* promoter which are a part of palindromic elements identified by Hennig et al. [25].

The above-mentioned transcription factors directly interact with the E-boxes of the E-cadherin gene promoter and inhibit E-cadherin expression when overexpressed for example in MDCK cells [24]. Downregulation of these transcription factors by siRNAs or antisense RNAs increases E-cadherin amounts and inhibits metastatic properties of many cancer cells [23].

In most epithelial cell types, members of the Snail family are effective inducers of EMT in embryogenesis and carcinogenesis. The overexpression of Snail1 stimulates EMT in epithelial cells thereby enhancing migration and invasion of the cells by reducing E-cadherin expression and enhancing mesenchymal-specific gene expression [26]. The diversity of the regulatory mechanisms as well as its dependence on the individual cellular situation is documented by the zinc finger proteins Snail and Slug. Snail and Slug are known to repress E-cadherin gene expression in a variety of cell types, but in mouse mammary NMuMG cells no significant impact on the activity of the E-cadherin promoter has been documented. In these cells, ZEB1 and ZEB2, also called SIP1 (Smad interacting protein 1), transcription factors are essentially required for inhibition of E-cadherin expression through interaction with E-box1 and E-box2 elements of the mouse E-cadherin promoter [27]. The zinc finger homeobox family members ZEB1 and SIP1 (ZEB2) are implicated as markers for EMT and malignant progression of various epithelial tumors [6]. ZEB1 directly influences the expression of the cell polarity factors AP1M2, PATJ, and CRB3 thereby inhibiting basal-apical differentiation and further promotes tumor progression [28, reviewed in 6]. SIP1 is upregulated in many epithelial tumors, such as gastric, ovary, esophageal, pancreatic or oral squamous cell carcinoma. Its expression has been correlated with reduced concentration of E-cadherin and induction of EMT [29–33]. In pancreatic tumor cells, SIP1 was identified as a mediator of ECM-modulated inhibition of intercellular adhesion, especially collagen type I-induced downregulation of E-cadherin. A Src-dependent upregulation of SIP1 correlated with reduced E-cadherin promoter activity in pancreatic tumor cells grown on collagen type I [33]. These data are supported by findings that in SIP1-knockout mouse embryos, E-cadherin was not downregulated in tissues, which normally express SIP1 in wild-type embryos, such as the neuroepithelium and the neural tube [34].

The transcriptional repressor Twist is a basic helix-loop-helix transcription factor and part of a signaling cascade that initiates mesoderm development during embryogenesis [35]. Upregulation of Twist induces EMT and metastasis including downregulation of E-cadherin especially in human breast epithelial cells [36] and lobular carcinoma [37]. Moreover, Twist does not only act as a repressor for E-cadherin but seems to be essential for the induction of mesenchymal proteins necessary for the metastasis of E-cadherin-deficient cells [38]. Based on the extensive work in characterizing the relation between all the above-mentioned transcription factors and E-cadherin, it has been shown that these are positioned upstream of E-cadherin. However, a recent report by Onder et al. [38] provides evidence that Twist and also ZEB1 is also upregulated in response to depletion of E-cadherin in mammary epithelia cells, indicating that the expression of these transcription factors may also be influenced by the level of E-cadherin, at least in some cells.

The Wnt-regulated transcription factor Lef-1 has also been identified as a regulator of E-cadherin gene expression. Jamora et al. [39] showed in mouse keratinocytes that Wnt-3a treatment resulted in transcriptionally competent Lef-1 complexes which bind to the E-cadherin promoter and inhibit its activity in luciferase reporter assays. These data are supported by a study of Nawshad et al. [40] showing that a transcriptionally active complex composed of phosphorylated Smad2, Smad4 and Lef-1 directly inhibits E-cadherin gene expression by binding to the *CDH1* promoter. Data from our own group further support this mode of E-cadherin regulation by showing that a Lef-1 splice variant lacking exon VI specifically interacts in cooperation with Smad proteins with the E-boxes of the human E-cadherin promoter and represses its activity [41].

The regulation of the discussed transcription factors themselves has not been elucidated sufficiently. Several studies point to the important role of the transforming growth factor-β (TGF-β) [42]. TGF-β has been reported to stimulate the expression of Snail and SIP1 mRNAs in various epithelial cells [23]. TGF-β-induced gene expression of Snail1 is induced via Smad3 in renal tubular epithelial cells [43] or via activation of the phosphatidylinositol-3 kinase (PI3-kinase) and the extracellular signal-regulated kinase (ERK) pathway in MDCK cells [44]. Transcriptional activation of SIP1 depends on its interaction with phosphorylated Smad2 to mediate E-cadherin gene expression [40]. Interestingly, SIP1 and ZEB1 expression is controlled by the transcription factor Ets1, another TGF-β target gene [45], which in turn is most likely controlled by Id2 (inhibitor of differentiation 2) [27]. The expression of Ets1 has been shown to repress E-cadherin expression in breast carcinoma and keratinocyte cell lines by binding to Ets-binding sites in the *CDH1* promoter region [46]. Proteins of the Id family are also targets of TGF-β-regulated gene expression and act through repression of E12/E47 transcription factors. Thus, downregulation of Id2 by TGF-β allows the inhibitory action of E12/E47 leading to E-cadherin repression and EMT [47]. Recently, two other regulators of Snail, Slug, SIP1 and Twist have been discovered as TGF-β targets. Inhibition of TGF-β receptor III (TβRIII) expression in mouse mammary epithelial cells results in increased activity of nuclear factor NFκB which in turn results in increased expression of the E-cadherin transcriptional repressors, especially of Snail [48]. Moreover, high mobility group factor A2 (HMGA2) is upregulated by TGF-β-induced binding of Smads to the HMGA2 promoter region in NMuMG mouse mammary cells. Forced expression of HMGA2 leads to upregulated expression of Snail, Slug and Twist expression and otherwise downregulation of Id2, which consequently results in marked inhibition of E-cadherin expression [49]. These findings strongly accentuate the relevance of TGF-β in induction of EMT and especially in the complex regulation of E-cadherin expression [more intensively reviewed in 50, 51].

Regulation of the E-Cadherin Adhesion Complex by β-Catenin

The fine modulation of the mechanical stability of the E-cadherin/catenin complex is mediated by a plethora of different and in parts unrelated molecular mechanisms. β-Catenin represents an important regulator of the E-cadherin/catenin-dependent intercellular adhesion, especially by posttranslational modification as well as part of Wnt/Wingless-induced signal transduction. The posttranslational modification of proteins by phosphorylation has been shown as an important instrument. Mainly, the phosphorylation of β-catenin at tyrosine residues has been suggested as an integral part of ECM- or growth factor-induced inhibition of the E-cadherin adhesion complex [5, 52]. The modification of tyrosine residues of catenins and especially of β-catenin has been shown to induce the dissociation of the E-cadherin/β-catenin/α-catenin adhesion complex. The binding of α-catenin to tyrosine-phosphorylated β-catenin is much weaker than to its unphosphorylated form resulting in diminished association of E-cadherin with the actin cytoskeleton. A strong interaction with the actin cytoskeleton represents one necessity for strong intercellular adhesion [52]. Three conserved tyrosine residues, namely Y142, Y489 and Y654, of β-catenin are involved in the interaction of β-catenin with other proteins especially α-catenin [reviewed by 52]. Tyrosine 142 is part of the α-catenin-binding domain of β-catenin and its phosphorylation results in the dissociation of α-catenin from β-catenin [53]. The phosphorylation of tyrosine 489 or tyrosine 654 by c-Abl or the epidermal growth factor receptor, respectively, has been associated with reduced affinity of β-catenin to cadherins [54]. The role of β-catenin phosphorylation in the regulation of the cell-cell adhesion complex is supported by the finding that expression of a chimeric E-cadherin/α-catenin fusion protein results in adherens junctions which are associated with the actin cytoskeleton but are insensitive to catenin phosphorylation [55, 56]. The increased tyrosine phosphorylation of β-catenin is mediated by different kinases such as the epidermal growth factor receptor, hepatocyte growth factor receptor c-met, Src or Fer [57] and the inhibition of phosphatases such as SHP2 (SH2 domain-containing inositol-5′-phosphatase 2), LAR, PTP or PTEN (phosphatase and tensin homolog) [58, 59]. Enhanced activities of growth factor receptors such as epidermal growth factor receptor or c-met or of cellular kinases such as c-Src, which are overexpressed and/or constitutively active in many tumors, are also responsible for phosphorylation of β-catenin. This enhanced phosphorylation is strongly correlated with carcinogenesis and metastasis in many different types of cancer [60].

A large body of evidence points towards a role of Src kinases in the regulation of cell-cell adhesion. Different reports describe the effect of activated Src on E-cadherin-mediated cellular adhesion suggesting a direct influence of activated Src on β-catenin phosphorylation followed by inhibition of cell aggregation [61–63]. Moreover, active Src regulates the cellular amount of E-cadherin. Activated Src phosphorylates E-cadherin at tyrosine residues which results in ubiquitylation by Hakai, a Cbl-like E3-ubiquitin ligase, and subsequent endocytosis and lysosomal degradation of E-cadherin [64]. Similar results were described for VE-cadherin, which is phosphorylated at Y658 and Y731 followed by its disappearance from adherens junctions of endothelial cells [65].

In addition to its mandatory role in cadherin-mediated cellular adhesion, β-catenin functions as a cotranscriptional factor involved in canonical Wnt signaling. A prerequisite for this function is the presence of a cytoplasmic pool of β-catenin which is regulated by the activity of the Wnt pathway. Stimulation by Wnt inhibits the activity of the glycogen synthase kinase-3β (GSK-3β). Active GSK-3β modifies β-catenin at serine/threonine residues (S33, S37 and T41) and induces its proteolytic degradation. Serine/threonine phosphorylation of β-catenin has not been associated with altered affinity of the E-cadherin/β-catenin/α-catenin complex but with regulation of the cytoplasmic β-catenin stability. The cytoplasmic localization of β-catenin is necessary for its nuclear translocation and its role in the regulation of gene expression mainly in cooperation with transcription factors of the Lef/TCF family [66]. Several genes have been identified to be regulated by β-catenin in cooperation with Lef/TCFs such as cyclin D1, c-myc, fibronectin and E-cadherin which are involved in processes during ontogenesis and carcinogenesis [38, described in detail in 66, 67].

Regulation of Cell-Cell Adhesion by the Cellular Microenvironment

Tumor cell invasion, the hallmark of malignant diseases, depends on the translocation of tumor cells from the initial neoplastic focus into neighboring host tissues and distant organs. In recent years it has become clear that this is not only a consequence of tumorigenesis but an active process at least partially controlled by the tumor cells themselves in producing a suitable environment which further promotes tumorigenesis and metastasis [68]. The tumor microenvironment is a complex system which changes dramatically during tumor formation. The microenvironment of tumor cells consists of the ECM, growth factors and different cell types, including not only tumor cells but also endothelial cells, smooth muscle cells, fibroblasts, myofibroblasts and immune cells. The ECM proteins are mainly produced and secreted by mesenchymal cells including stromal fibroblasts, cancer-associated fibroblast, activated myofibroblast or activated stellate cells [69]. This specific microenvironment influences the cellular behavior of the tumor cells by providing cytokines, growth factors and proteases which promote chemotaxis and invasion [70]. Signaling processes initiated by interactions between tumor cells and their surrounding induce EMT and regulate E-cadherin localization and function. In some carcinoma, the membranous localization of E-cadherin is lost only in undifferentiated areas of the tumor, particularly near the invasive front as shown for colon, squamous cell carcinoma and some pancreatic tumors [6]. This observation is most likely explained by the influence of the microenvironment which is different at the invasive front compared to central areas of the tumor and compared to metastases which contains a totally different environment often with less stroma [71]. In most colon carcinoma, the ECM near the invasive front contains much more laminin 5 as compared to central parts of the tumor, which has suggested to be a proinvasive factor for colon carcinoma [72]. Thus, the specific composition of the ECM might be the trigger by which the environment determines characteristics of the tumor cells.

In recent years a great number of studies have emphasized the importance of the composition of the microenvironment. The composition changes during tumor progression from cross-linked collagens, such as collagen type IV present in the basal membrane, to fibrillar collagen, such as collagen type I or III in tumor stroma and is accompanied by the presence of different growth factors. In addition, several reports highlight the interplay between cancer cells and mesenchymal cells in the tumor environment, like cancer-associated fibroblast or myofibroblast, which adds to the complex regulatory network influencing tumor growth and invasion. During development of the mammary gland and of breast cancer, the important influence of mesenchymal cell on the differentiation of epithelial cells has been studied intensively [71, 73]. The activation of these mesenchymal cells during tumorigenesis is often caused by growth factors, such as TGF-β, platelet-derived growth factor, fibroblast growth factor or cytokines released by the tumor cells themselves. Mesenchymal stellate cells, which are especially described in liver and pancreas [74], synthesize

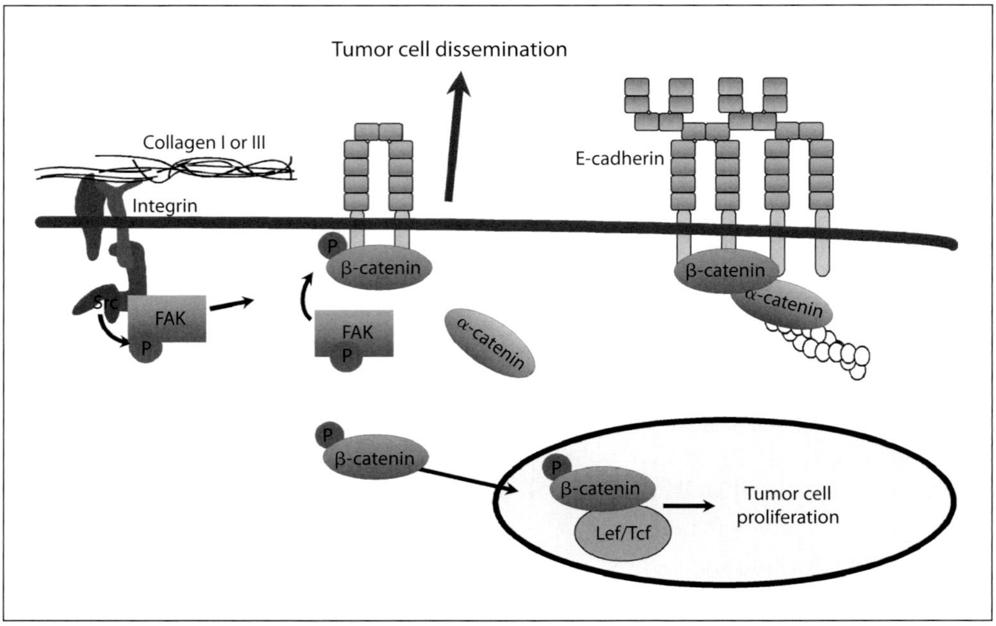

Fig. 2. Molecular details of the potential crosstalk between integrin-dependent cell-matrix adhesion and E-cadherin-mediated cell-cell adhesion.

and secret excessive amounts of ECM proteins [74]. The resulting stroma contains mainly fibrillar type I and III collagens, fibronectin, laminin and glycoproteins which stimulates tumor cell growth, angiogenesis and invasion [reviewed in 75, 76]. The cellular receptors for most ECM proteins are members of the integrin superfamily, which are transmembrane α/β heterodimeric complexes. Integrins are activated in response to binding to ECM proteins and initiate a plethora of downstream signal transduction cascades [77]. One example relevant to tumor invasion is the initiation of cell-substrate adhesion by the assembly of hemidesmosomes or focal contacts. Integrin ligation represents a central part in the assembly of focal contacts by attracting structural proteins such as talin, paxillin or vinculin, thereby activating multiple downstream signaling molecules including the kinases Src and focal adhesion kinase (FAK) [78, 79]. FAK and Src are phosphorylated and thus activated within minutes after collagen-induced integrin activation, which is accompanied by the formation of focal adhesion complexes [5, 79]. Due to its multiple interaction partners, such as members of the p130/Crk/DOCK1 cascade, the Raf-MEK-ERK pathway as well as members of the PI3-kinase-Akt cascade [79], the non-receptor tyrosine kinase FAK seems to be a major player in collagen-induced signal transduction. Most interestingly, we have demonstrated that collagen type I stimulation of pancreatic carcinoma cells induces integrin-dependent activation of FAK and translocation of activated FAK to E-cadherin complexes in the apical part of differentiated epithelial pancreatic cells. This presence of activated FAK at the E-cadherin complex is associated with an enhanced tyrosine phosphorylation of β-catenin which results in a disassembly of the E-cadherin/catenin complexes and subsequently weakened cell-cell interactions [55]. In colon cancer cells, expression of FAK mutants, which were resistant to Src kinase-mediated phosphorylation, stabilize the E-cadherin-mediated cell-cell adhesion [80]. Moreover, expression of dominant-negative FAK or inhibition of Src in Src-overexpressing epithelial colon cancer cells restored E-cadherin-mediated cellular adhesion [80]. In summary, these findings highlight the role of FAK in EMT by modulating E-cadherin-dependent cell-cell adhesion and pointing to a crosstalk between integrin-mediated cell-substrate adhesion and E-cadherin-mediated cell-cell adhesion in epithelial tumors [33, 52, 55]. Figure 2 briefly summarizes this hypothesis regarding the crosstalk between integrins and E-cadherin.

Another model of integrin signaling in the regulation of adherens junctions underlines the role of talin. The scaffold protein talin contains multiple interaction domains for adaptor proteins, kinases and phosphatases and links activated integrins to signaling pathways involving Src family kinases, the Ras-ERK or the Rho GTPase cascade [79, 81]. As mentioned before, activation of Src kinase through ligation of αvβ3 integrins has been shown to mediate VE-cadherin phosphorylation in endothelial cells, which leads to dissociation of β-catenin and p120ctn. Subsequently VE-cadherin disappeared from cell-cell contacts thereby reducing intercellular adhesion [65].

Regulation of Cell-Cell Adhesion by p120ctn and GTPases

The catenin p120 (p120ctn) is likewise believed to play a pivotal role in the regulation of different aspects of E-cadherin/catenin adhesion complexes. One important function of p120ctn is its ability to interact with the microfilament system. p120ctn promotes cell surface trafficking of E-cadherin by binding to kinesin and transportation of E-cadherin along microtubules towards the plasma membrane [82]. Studies on the function of p120ctn in epithelial cells revealed that the protein controls the availability of E-cadherin by regulating the recycling of endocytosed molecules [reviewed in 83, 84]. While E-cadherin is rapidly degraded in cells with low p120ctn content, forced expression of p120ctn prevents the lysosomal degradation of E-cadherin [85]. Inhibition of lysosomal acidification also prevents E-cadherin from degradation in p120ctn-deficient cells [86]. The inducible expression of p120ctn in cells with low E-cadherin level can restore E-cadherin protein concentration and prevents E-cadherin redistribution into the cytoplasm, which subsequently results in a reinforcement of cell-cell adhesion [84]. Thus, it has been suggested that the E-cadherin-bound p120ctn acts as a protector against cadherin modification which would mark E-cadherin for internalization and destruction [84].

The complex role of p120ctn in epithelial cells is underscored by the observations that p120ctn increases the ability of cadherin molecules for lateral clustering. In the

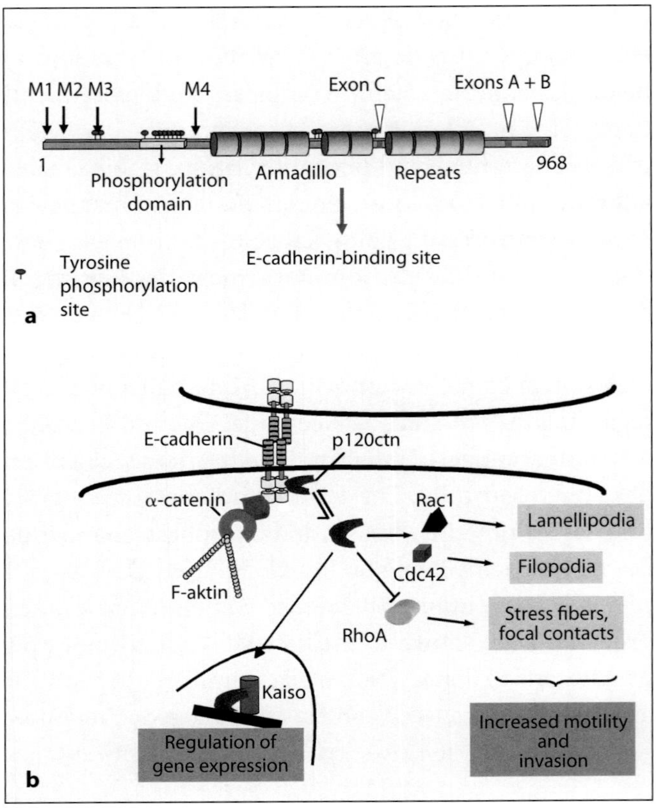

Fig. 3. Role of catenin p120 (p120ctn) in the regulation of cell-cell adhesion and cell motility. **a** Schematic presentation of p120ctn protein structure. **b** Role of p120ctn in the assembly of the E-cadherin adhesion complex, the regulation of gene expression and the control of cell migration.

presence of p120ctn, E-cadherin molecules form *cis*-dimers and build a zipper-like adherens junctions thereby strengthening cell-cell adhesion [87]. To make it even more complex, p120ctn exists in multiple isoforms initiated from at least four different start codons and three alternatively spliced exons which result in a huge number of possible isoforms (fig. 3). Although the exact role of the different p120ctn isoforms has not been fully characterized yet, longer isoforms initiated from start codon 1 or 2 are preferentially expressed in differentiated epithelial cells whereas shorter isoforms starting from start codon 3 or 4 are present in mesenchymal, migrating cells as well as several carcinoma-derived cell lines [88]. In addition to differences in the p120ctn isoforms, phosphorylation at different tyrosine residues has been identified. These phosphorylation sites, which are mainly localized at the N-terminal part of p120ctn in isoforms 1 and 2 only, modulate the binding to cadherins and influence the stability of E-cadherin/catenin complexes [89]. Besides directly influencing E-cadherin protein content and cell-cell adhesion, p120ctn regulates the transcription factor Kaiso in controlling gene expression [90]. p120ctn has been detected inside the nucleus in some cell lines. Indeed, p120ctn possesses classical nuclear localization sites and contributes to the regulation of the DNA-binding and transcriptional activity of the transcription factor Kaiso [90].

There is strong evidence that p120ctn acts, at least partially, by controlling Rho GTPases. Different reports show that p120ctn inhibits the activity of RhoA most likely by binding to the GTPases activating protein p190RhoGAP. In addition, p120ctn contributes to the activation of Rac1 and Cdc42 [91, 92]. Forced expression of p120ctn in fibroblasts or cadherin-deficient epithelial cells produces a typical morphology called the 'branching phenotype' which is characterized by arborization of cellular protrusions [93]. This crosstalk suggests a number of plausible mechanisms through which p120ctn could promote cell-cell adhesion or cell motility [93] (summarized in figure 3). Although the molecular mechanisms responsible for p120ctn-Rho-GTPases-induced signaling events needs to be established, it has been shown that localized restrictions between Rac1 and RhoA activation accompanies successful formation of stable cell-cell contacts [94]. Rac1 activation was found in lamellipodia initially localized to new cell-cell contacts. The subsequent E-cadherin accumulation correlates with diminished Rac1 activity. In extending cell-cell contacts, active RhoA was described only at the edge of growing cell-cell contacts, where it is necessary to drive expansion and completion of cell-cell adhesion [94]. RhoA activity seems to be repressed in the center of the growing adherens junction, which may be induced by p120ctn attraction to the growing adherens junctions.

Although a great quantity of data has been collected in recent years about the molecular mechanisms leading to dissemination of tumor cells, metastasis and invasion, future progress is necessary to understand these processes in detail. It will be of great interest to improve the understanding of the involvement and interdependence of the cadherin-mediated adherens junctions with other cell adhesion modules, the actin cytoskeleton and regulatory proteins such as the Rho GTPases in the formation and maintenance of intercellular adhesion in embryogenesis and most important in tumor development.

References

1 Hahn WC, Weinberg RA: Modelling the molecular circuitry of cancer. Nat Rev Cancer 2002;2:331–341.

2 Brooks SA, Lomax-Browne HJ, Carter TM, Kinch CE, Hall DM: Molecular interactions in cancer cell metastasis. Acta Histochem 2009;doi: 10.1016/j.acthis.2008.11.002.

3 Chiang AC, Massague J: Molecular basis of metastasis. N Engl J Med 2008;359:2814–2823.

4 Cavallaro U, Niedermeyer J, Fuxa M, Christofori G: N-CAM modulates tumour-cell adhesion to matrix by inducing FGF-receptor signalling. Nat Cell Biol 2001;3:650–657.

5 Menke A, Philippi C, Vogelmann R, Seidel B, Lutz MP, Adler G, Wedlich D: Down-regulation of E-cadherin gene expression by collagen type I and type III in pancreatic cancer cell lines. Cancer Res 2001;61:3508–3517.

6 Schmalhofer O, Brabletz S, Brabletz T: E-cadherin, β-catenin, and ZEB1 in malignant progression of cancer. Cancer Metastasis Rev 2009;28:151–166.

7 Yang J, Weinberg RA: Epithelial-mesenchymal transition: at the crossroads of development and tumor metastasis. Dev Cell 2008;14:818–829.

8 Usami Y, Satake S, Nakayama F, Matsumoto M, Ohnuma K, Komori T, Semba S, Ito A, Yokozaki H: Snail-associated epithelial-mesenchymal transition promotes oesophageal squamous cell carcinoma motility and progression. J Pathol 2008;215:330–339.

9 Gunji N, Oda T, Todoroki T, Kanazawa N, Kawamoto T, Yuzawa K, Scarpa A, Fukao K: Pancreatic carcinoma: correlation between E-cadherin and α-catenin expression status and liver metastasis. Cancer 1998;82:1649–1656.

10 Van Roy F, Berx G: The cell-cell adhesion molecule E-cadherin. Cell Mol Life Sci 2008;65:3756–3788.
11 Vandewalle C, van Roy F, Berx G: The role of the ZEB family of transcription factors in development and disease. Cell Mol Life Sci 2009;66:773–787.
12 Niessen CM: Tight junctions/adherens junctions: basic structure and function. J Invest Dermatol 2007; 127:2525–2532.
13 Dupin I, Camand E, Etienne-Manneville S: Classical cadherins control nucleus and centrosome position and cell polarity. J Cell Biol 2009;185:779–786.
14 Kemler R: From cadherins to catenins: cytoplasmic protein interactions and regulation of cell adhesion. Trends Genet 1993;9:317–321.
15 Nelson WJ: Regulation of cell-cell adhesion by the cadherin-catenin complex. Biochem Soc Trans 2008; 36:149–155.
16 Hinck L, Näthke IS, Papkoff J, Nelson WJ: β-Catenin: a common target for the regulation of cell adhesion by Wnt-1 and Src signaling pathways. Trends Biochem Sci 1994;19:538–542.
17 Huber MA, Kraut N, Beug H: Molecular requirements for epithelial-mesenchymal transition during tumor progression. Curr Opin Cell Biol 2005;17: 548–558.
18 Thiery JP: Epithelial-mesenchymal transitions in tumour progression. Nat Rev Cancer 2002;2:442–454.
19 Mareel M, Boterberg T, Noe V, Van Hoorde L, Vermeulen S, Bruyneel E, Bracke M: E-cadherin/catenin/cytoskeleton complex: a regulator of cancer invasion. J Cell Physiol 1997;173:271–274.
20 Hirohashi S, Kanai Y: Cell adhesion system and human cancer morphogenesis. Cancer Sci 2003;94: 575–581.
21 Nass SJ, Herman JG, Gabrielson E, Iversen PW, Parl FF, Davidson NE, Graff JR: Aberrant methylation of the estrogen receptor and E-cadherin 5′ CpG islands increases with malignant progression in human breast cancer. Cancer Res 2000;60:4346–4348.
22 Becker KF, Atkinson MJ, Reich U, Huang HH, Nekarda H, Siewert JR, Hofler H: Exon skipping in the E-cadherin gene transcript in metastatic human gastric carcinomas. Hum Mol Genet 1993;2:803–804.
23 Peinado H, Olmeda D, Cano A: Snail, ZEB and bHLH factors in tumour progression: an alliance against the epithelial phenotype? Nat Rev Cancer 2007;7:415–428.
24 Moreno-Bueno G, Cubillo E, Sarrio D, Peinado H, Rodriguez-Pinilla SM, Villa S, Bolos V, Jorda M, Fabra A, Portillo F, Palacios J, Cano A: Genetic profiling of epithelial cells expressing E-cadherin repressors reveals a distinct role for Snail, Slug, and E47 factors in epithelial-mesenchymal transition. Cancer Res 2006;66:9543–9556.
25 Hennig G, Behrens J, Truss M, Frisch S, Reichmann E, Birchmeier W: Progression of carcinoma cells is associated with alterations in chromatin structure and factor binding at the E-cadherin promoter in vivo. Oncogene 1995;11:475–484.
26 De Wever O, Pauwels P, De Craene B, Sabbah M, Emami S, Redeuilh G, Gespach C, Bracke M, Berx G: Molecular and pathological signatures of epithelial-mesenchymal transitions at the cancer invasion front. Histochem Cell Biol 2008;130:481–494.
27 Shirakihara T, Saitoh M, Miyazono K: Differential regulation of epithelial and mesenchymal markers by δEF1 proteins in epithelial mesenchymal transition induced by TGF-β. Mol Biol Cell 2007;18:3533–3544.
28 Spaderna S, Schmalhofer O, Wahlbuhl M, Dimmler A, Bauer K, Sultan A, Hlubek F, Jung A, Strand D, Eger A, Kirchner T, Behrens J, Brabletz T: The transcriptional repressor ZEB1 promotes metastasis and loss of cell polarity in cancer. Cancer Res 2008;68: 537–544.
29 Castro AC, Rosivatz E, Schott C, Hollweck R, Becker I, Sarbia M, Carneiro F, Becker KF: Slug is overexpressed in gastric carcinomas and may act synergistically with SIP1 and Snail in the down-regulation of E-cadherin. J Pathol 2007;211:507–515.
30 Maeda G, Chiba T, Okazaki M, Satoh T, Taya Y, Aoba T, Kato K, Kawashiri S, Imai K: Expression of SIP1 in oral squamous cell carcinomas: implications for E-cadherin expression and tumor progression. Int J Oncol 2005;27:1535–1541.
31 Elloul S, Silins I, Trope CG, Benshushan A, Davidson B, Reich R: Expression of E-cadherin transcriptional regulators in ovarian carcinoma. Virchows Arch 2006; 449:520–528.
32 Rees JR, Onwuegbusi BA, Save VE, Alderson D, Fitzgerald RC: In vivo and in vitro evidence for transforming growth factor-β_1-mediated epithelial to mesenchymal transition in esophageal adenocarcinoma. Cancer Res 2006;66:9583–9590.
33 Imamichi Y, König A, Gress T, Menke A: Collagen type I-induced Smad-interacting protein 1 expression downregulates E-cadherin in pancreatic cancer. Oncogene 2007;26:2381–2385.
34 Van de Putte T, Maruhashi M, Francis A, Nelles L, Kondoh H, Huylebroeck D, Higashi Y: Mice lacking ZFHX1B, the gene that codes for Smad-interacting protein-1, reveal a role for multiple neural crest cell defects in the etiology of Hirschsprung disease-mental retardation syndrome. Am J Hum Genet 2003;72:465–470.
35 Castanon I, Baylies MK: A Twist in fate: evolutionary comparison of Twist structure and function. Gene 2002;287:11–22.

36 Kang Y, Massague J: Epithelial-mesenchymal transitions: twist in development and metastasis. Cell 2004;118:277–279.

37 Yang J, Mani SA, Donaher JL, Ramaswamy S, Itzykson RA, Come C, Savagner P, Gitelman I, Richardson A, Weinberg RA: Twist, a master regulator of morphogenesis, plays an essential role in tumor metastasis. Cell 2004;117:927–939.

38 Onder TT, Gupta PB, Mani SA, Yang J, Lander ES, Weinberg RA: Loss of E-cadherin promotes metastasis via multiple downstream transcriptional pathways. Cancer Res 2008;68:3645–3654.

39 Jamora C, DasGupta R, Kocieniewski P, Fuchs E: Links between signal transduction, transcription and adhesion in epithelial bud development. Nature 2003;422:317–322.

40 Nawshad A, Medici D, Liu CC, Hay ED: TGF-β_3 inhibits E-cadherin gene expression in palate medial-edge epithelial cells through a Smad2-Smad4-LEF1 transcription complex. J Cell Sci 2007;120:1646–1653.

41 Jesse S, Koenig A, Ellenrieder V, Menke A: Lef-1 isoforms regulate different target genes and reduce cellular adhesion. Int J Cancer 2009;130:doi: 10.1002/ijc.24802.

42 Xu J, Lamouille S, Derynck R: TGF-β-induced epithelial to mesenchymal transition. Cell Res 2009;19:156–172.

43 Sato M, Muragaki Y, Saika S, Roberts AB, Ooshima A: Targeted disruption of TGF-β_1/Smad3 signaling protects against renal tubulointerstitial fibrosis induced by unilateral ureteral obstruction. J Clin Invest 2003;112:1486–1494.

44 Peinado H, Quintanilla M, Cano A: Transforming growth factor-β_1 induces snail transcription factor in epithelial cell lines: mechanisms for epithelial mesenchymal transitions. J Biol Chem 2003;278:21113–21123.

45 Xie L, Law BK, Aakre ME, Edgerton M, Shyr Y, Bhowmick NA, Moses HL: Transforming growth factor β-regulated gene expression in a mouse mammary gland epithelial cell line. Breast Cancer Res 2003;5:R187–R198.

46 Rodrigo I, Cato AC, Cano A: Regulation of E-cadherin gene expression during tumor progression: the role of a new Ets-binding site and the E-pal element. Exp Cell Res 1999;248:358–371.

47 Kondo M, Cubillo E, Tobiume K, Shirakihara T, Fukuda N, Suzuki H, Shimizu K, Takehara K, Cano A, Saitoh M, Miyazono K: A role for Id in the regulation of TGF-β-induced epithelial-mesenchymal transdifferentiation. Cell Death Differ 2004;11:1092–1101.

48 Criswell TL, Arteaga CL: Modulation of NFκB activity and E-cadherin by the type III transforming growth factor β receptor regulates cell growth and motility. J Biol Chem 2007;282:32491–32500.

49 Thuault S, Tan EJ, Peinado H, Cano A, Heldin CH, Moustakas A: HMGA2 and Smads co-regulate SNAIL1 expression during induction of epithelial-to-mesenchymal transition. J Biol Chem 2008;283:33437–33446.

50 Padua D, Massague J: Roles of TGF-β in metastasis. Cell Res 2009;19:89–102.

51 Giehl K, Imamichi Y, Menke A: Smad4-independent TGF-β signaling in tumor cell migration. Cells Tissues Organs 2007;185:123–130.

52 Lilien J, Balsamo J: The regulation of cadherin-mediated adhesion by tyrosine phosphorylation/dephosphorylation of β-catenin. Curr Opin Cell Biol 2005;17:459–465.

53 Piedra J, Miravet S, Castaño J, Pálmer HG, Heisterkamp N, García de Herreros A, Duñach M: p120 catenin-associated Fer and Fyn tyrosine kinases regulate β-catenin Tyr-142 phosphorylation and β-catenin-α-catenin Interaction. Mol Cell Biol 2003;23:2287–2297.

54 Huber AH, Weis WI: The structure of the β-catenin/E-cadherin complex and the molecular basis of diverse ligand recognition by β-catenin. Cell 2001;105:391–402.

55 Koenig A, Mueller C, Hasel C, Adler G, Menke A: Collagen type I induces disruption of E-cadherin-mediated cell-cell contacts and promotes proliferation of pancreatic carcinoma cells. Cancer Res 2006;66:4662–4671.

56 Nagafuchi A, Ishihara S, Tsukita S: The roles of catenins in the cadherin-mediated cell adhesion: functional analysis of E-cadherin-α catenin fusion molecules. J Cell Biol 1994;127:235–245.

57 Roura S, Miravet S, Piedra J, García de Herreros A, Duñach M. Regulation of E-cadherin/catenin association by tyrosine phosphorylation. J Biol Chem 1999;274:36734–36740.

58 Vogelmann R, Nguyen-Tat MD, Giehl K, Adler G, Wedlich D, Menke A: TGF-β-induced downregulation of E-cadherin-based cell-cell adhesion depends on PI3-kinase and PTEN. J Cell Sci 2005;118:4901–4912.

59 Müller T, Choidas A, Reichmann E, Ullrich A: Phosphorylation and free pool of β-catenin are regulated by tyrosine kinases and tyrosine phosphatases during epithelial cell migration. J Biol Chem 1999;274:10173–10183.

60 Mareel M, Leroy A: Clinical, cellular, and molecular aspects of cancer invasion. Physiol Rev 2003;83:337–376.

61 Imamichi Y, Menke A: Signaling pathways involved in collagen-induced disruption of the E-cadherin complex during epithelial-mesenchymal transition. Cells Tissues Organs 2007;185:180–190.
62 Koenig A, Mueller C, Hasel C, Adler G, Menke A: Collagen type I induces disruption of E-cadherin-mediated cell-cell contacts and promotes proliferation of pancreatic carcinoma cells. Cancer Res 2006; 66:4662–4671.
63 Hamaguchi M, Matsuyoshi N, Ohnishi Y, Gotoh B, Takeichi M, Nagai Y: p60v-src causes tyrosine phosphorylation and inactivation of the N-cadherin-catenin cell adhesion system. EMBO J 1993;12: 307–314.
64 Fujita Y, Krause G, Scheffner M, Zechner D, Leddy HE, Behrens J, Sommer T, Birchmeier W: Hakai, a c-Cbl-like protein, ubiquitinates and induces endocytosis of the E-cadherin complex. Nat Cell Biol 2002; 4:222–231.
65 Alghisi GC, Ponsonnet L, Ruegg C: The integrin antagonist cilengitide activates αVβ3, disrupts VE-cadherin localization at cell junctions and enhances permeability in endothelial cells. PLoS One 2009;4:e4449.
66 Clevers H: Wnt/β-catenin signaling in development and disease. Cell 2006;127:469–480.
67 Hoppler S, Kavanagh CL: Wnt signalling: variety at the core. J Cell Sci 2007;120:385–393.
68 Albini A, Sporn MB: The tumour microenvironment as a target for chemoprevention. Nat Rev Cancer 2007;7:139–147.
69 Mahadevan D, Von Hoff DD: Tumor-stroma interactions in pancreatic ductal adenocarcinoma. Mol Cancer Ther 2007;6:1186–1197.
70 Moustakas A, Heldin CH: Signaling networks guiding epithelial-mesenchymal transitions during embryogenesis and cancer progression. Cancer Sci 2007;98:1512–1520.
71 De Wever O, Mareel M: Role of tissue stroma in cancer cell invasion. J Pathol 2003;200:429–447.
72 Hlubek F, Spaderna S, Jung A, Kirchner T, Brabletz T: Beta-catenin activates a coordinated expression of the proinvasive factors laminin-5 γ2 chain and MT1-MMP in colorectal carcinomas. Int J Cancer 2004;108:321–326.
73 Kalluri R, Zeisberg M: Fibroblasts in cancer. Nat Rev Cancer 2006;6:392–401.
74 Kleeff J, Beckhove P, Esposito I, Herzig S, Huber PE, Lohr JM, Friess H: Pancreatic cancer microenvironment. Int J Cancer 2007;121:699–705.
75 Hay ED: Extracellular matrix alters epithelial differentiation. Curr Opin Cell Biol 1993;5:1029–1035.
76 Angeli F, Koumakis G, Chen MC, Kumar S, Delinassios JG: Role of stromal fibroblasts in cancer: promoting or impeding? Tumour Biol 2009;30:109–120.
77 Berrier AL, Yamada KM: Cell-matrix adhesion. J Cell Physiol 2007;213:565–573.
78 Fashena SJ, Thomas SM: Signalling by adhesion receptors. Nat Cell Biol 2000;2:E225–E229.
79 Mitra SK, Schlaepfer DD: Integrin-regulated FAK-Src signaling in normal and cancer cells. Curr Opin Cell Biol 2006;18:516–523.
80 Avizienyte E, Wyke AW, Jones RJ, McLean GW, Westhoff MA, Brunton VG, Frame MC: Src-induced de-regulation of E-cadherin in colon cancer cells requires integrin signalling. Nat Cell Biol 2002;4: 632–638.
81 Brakebusch C, Fässler R: The integrin-actin connection, an eternal love affair. EMBO J 2003;22: 2324–2333.
82 Stehbens SJ, Paterson AD, Crampton MS, Shewan AM, Ferguson C, Akhmanova A, Parton RG, Yap AS: Dynamic microtubules regulate the local concentration of E-cadherin at cell-cell contacts. J Cell Sci 2006;119:1801–1811.
83 Reynolds AB, Roczniak-Ferguson A: Emerging roles for p120-catenin in cell adhesion and cancer. Oncogene 2004;23:7947–7956.
84 Kowalczyk AP, Reynolds AB. Protecting your tail: regulation of cadherin degradation by p120-catenin. Curr Opin Cell Biol 2004;16:522–527.
85 Ireton RC, Davis MA, van Hengel J, Mariner DJ, Barnes K, Thoreson MA, Anastasiadis PZ, Matrisian L, Bundy LM, Sealy L, Gilbert B, van Roy F, Reynolds AB: A novel role for p120 catenin in E-cadherin function. J Cell Biol 2002;159:465–476.
86 Xiao K, Allison DF, Buckley KM, Kottke MD, Vincent PA, Faundez V, Kowalczyk AP: Cellular levels of p120 catenin function as a set point for cadherin expression levels in microvascular endothelial cells. J Cell Biol 2003;163:535–545.
87 Thoreson MA, Anastasiadis PZ, Daniel JM, Ireton RC, Wheelock MJ, Johnson KR, Hummingbird DK, Reynolds AB: Selective uncoupling of p120(ctn) from E-cadherin disrupts strong adhesion. J Cell Biol 2000;148:189–202.
88 Mayerle J, Friess H, Büchler M, Schnekenburger J, Weiss F, Klaus P, Domschke W, Lerch M: Up-regulation, nuclear import, and tumor growth stimulation of the adhesion protein p120(ctn) in pancreatic cancer. Gastroenterology 2003;124:949–960.
89 Seidel B, Braeg S, Adler G, Wedlich D, Menke A: E- and N-cadherin differ with respect to their associated p120ctn isoforms and their ability to suppress invasive growth in pancreatic cancer cells. Oncogene 2004;23:5532–5542.
90 Daniel JM: Dancing in and out of the nucleus: p120(ctn) and the transcription factor Kaiso. Biochim Biophys Acta 2007;1773:59–68.

91 Noren NK, Liu BP, Burridge K, Kreft B: p120 catenin regulates the actin cytoskeleton via Rho family GTPases. J Cell Biol 2000;150:567–580.

92 Wildenberg GA, Dohn MR, Carnahan RH, Davis MA, Lobdell NA, Settleman J, Reynolds AB: p120-catenin and p190RhoGAP regulate cell-cell adhesion by coordinating antagonism between Rac and Rho. Cell 2006;127:1027–1039.

93 Anastasiadis PZ: p120-ctn: a nexus for contextual signaling via Rho GTPases. Biochim Biophys Acta 2007;1773:34–46.

94 Yamada S, Nelson WJ: Localized zones of Rho and Rac activities drive initiation and expansion of epithelial cell-cell adhesion. J Cell Biol 2007;178:517–527.

Dr. Klaudia Giehl
Department of Internal Medicine I, University of Ulm
DE–89070 Ulm (Germany)
Tel. +49 731 500 44682, Fax +49 731 500 44502, E-Mail klaudia.giehl@uni-ulm.de

Entschladen F, Zänker KS (eds): Cell Migration: Signalling and Mechanisms.
Transl Res Biomed. Basel, Karger, 2010, vol 2, pp 136–162

The Cytoskeletal Connection: Understanding Adaptor Proteins

Wolfgang H. Ziegler

Interdisciplinary Center for Clinical Research (IZKF) Leipzig, Faculty of Medicine, University of Leipzig, Leipzig, Germany

Abstract

The cytoskeletal connection of integrin-based cell contacts is crucial to coordinated cell motility and migration, linking actin-mediated changes of cell shape to receptor-based adhesion. The connection allows transmission of forces to the cellular environment, sensing of chemical, topological and mechanical information, and retraction of structures as required by the cell. The interface between transmembrane receptors of the integrin family and the actin cytoskeleton critically depends on self-organized large multi-protein complexes, so-called cell adhesion complexes. The molecular architecture of these complexes comprising up to 100 or even more different proteins is very dynamic and allows control of turnover, as well as maturation and remodeling of adhesion sites to specific cellular needs. Given the complexity and versatility of the system, rules governing the assembly, maturation and dissociation of cell adhesion complexes are only beginning to be understood. Although many constituents of cell adhesion complexes were purified and sequenced nearly two decades ago, structural properties and biological function of cytoskeletal adaptor proteins are still poorly understood or were elucidated to greater detail only in recent years. The scope of this article is to discuss concepts and implications of analyses performed on two pivotal cytoskeletal proteins, talin and vinculin, and their relation to the current understanding of integrin-mediated cell adhesion. Diverse functional aspects, reaching from molecular structures of the proteins to functional integrity of organs, will be discussed.

The Cell Adhesion Complex: A Simplistic View

Cell contacts are not limited to providing anchor points that allow force transmission to selected extracellular ligands. Organization and signaling of adhesion sites rather supply cells with information on the chemical nature, topology and spacing of extracellular ligands in addition to the compliance of the cellular environment. Acquisition, processing, and transmission of these different informational cues critically depend on the interaction of the proteins building the cell adhesion complex. Although many components and interactions are known, a comprehensive, functional model of the cell adhesions complex is far from completion [1–3]. For the purpose of this discussion, a very basic model of integrin-based adhesion structures will be employed to illustrate fundamental concepts of protein function (fig. 1). In this model, the transmembrane receptor, an αβ-heterodimer of the integrin family, is connected externally to protein fibers of the extracellular matrix or equivalently to a transmembrane ligand of a neighboring cell. Internally, the integrin receptor is connected to filaments of the actin cytoskeleton (F-actin). Cytoskeletal association of the receptor is mediated by adaptor proteins, talin and vinculin, which convey essential functions of the cell adhesion complex. They modulate adhesion site turnover and allow regulation of F-actin binding and force transmission.

To understand the basic function and self-organization of the system, some key characteristics of the protein components have to be considered: (1) The integrin receptor acquires different states of activity and can be activated by binding partners from the inside and/or the outside of the cell. The best characterized internal binding partner mediating activation is talin. Binding of talin to β-integrins is proposed to re-orientate αβ-integrin binding leading to separation of their short cytodomains and increased ligand binding affinity of the extracellular domains. (2) The cytoskeletal proteins talin and vinculin perform autoinhibitory intramolecular interactions, which effectively block ligand binding outside of adhesion sites. Both proteins become activated in cell contacts, where they interact simultaneously with different binding partners. Competition between ligand binding and inhibitory intramolecular interaction enables the cytoskeletal proteins to shuttle dynamically between their 'inactive' cytoplasmic and their 'active', ligand-bound conformation in adhesion sites. Dynamic association of adaptor protein is assumed to form the basis of cell adhesion complex remodeling and turnover. (3) The cytoskeletal adaptor proteins provide furthermore flexible and functionally different contact sites for individual filaments and bundles of F-actin. Being organized in bundles and networks as well as large contractile actomyosin fibers (stress fibers), actin filaments turn over constantly and provide very flexible structures that form the basis for active changes of the cell's shape, pushing out of extensions and retraction. Together, these simple characteristics outline the central challenge and duty of the cell adhesion complex, which is the attachment of a relatively static anchor point, the transmembrane receptor patch, to a sterically and functionally dynamic force generator, the actin cytoskeleton. (4) Force appears to be one if not the most important organizing principle. Stability and remodeling of the cell adhesion complex critically depend on the transmission of force across the cytoskeletal connection. When the interaction is uncoupled on either side, e.g. at the outside by displacement of receptor ligands or inside the cells by inhibition or disruption of contractile actomyosin fibers, the entire complex disassembles and the cell contact dissolves. Furthermore, (5) cell adhesion sites also play a central role in the modulation of intracellular signaling. Cytoskeletal proteins in their 'active' conformation typically contain multiple ligand-binding sites, which can provide scaffold function for signal transduction pathways. Consistently, cell contact sites are highly

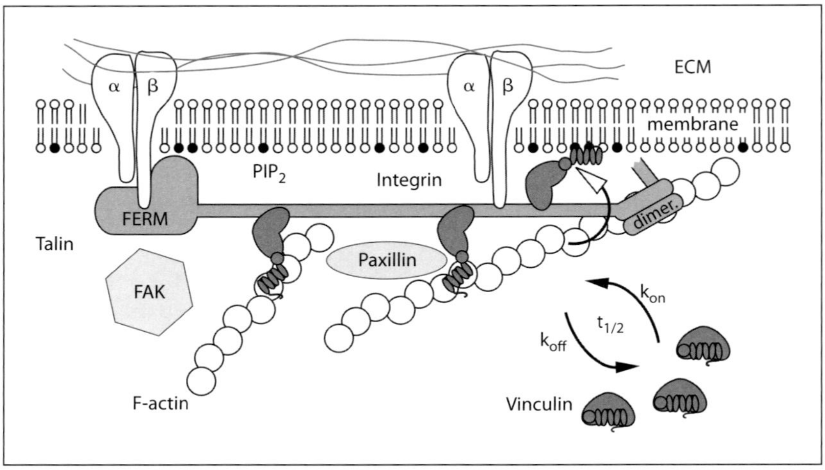

Fig. 1. Integrin-based cell adhesion: selected interactions of adaptor proteins talin and vinculin. In this basic model, integrin αβ-heterodimers are bound to extracellular matrix (ECM). Inside the cell, β-integrin cytodomains can bind to talin head (FERM) and rod (dimer, talin dimer is indicated). Together with talin, vinculin provides a mechanical connection to the actin cytoskeleton (F-actin). Interactions of the vinculin head with several sites in talin rod and of vinculin tail with F-actin and the plasma membrane are shown. FAK and paxillin are examples of alternative binding partners of talin and vinculin, respectively that are involved in adhesion site signalling. For vinculin, dynamic shuttling between the autoinhibited, cytoplasmic and the 'active', ligand-bound conformation is exemplified (arrow pair, k_{on}/k_{off}). Competition of actin binding by PIP_2 (black dots) is indicated (open arrow).

enriched in signaling molecules and phosphorylated proteins. In our simple model, this aspect is exemplified by (alternative) binding partners paxillin and focal adhesion kinase (FAK) of vinculin and talin, respectively.

Using talin and vinculin as a paradigm for functional interaction of cytoskeletal adaptor proteins, their properties relevant to the function of the cytoskeletal connection will be discussed at the different organizational levels from molecular interaction of proteins to specific requirements of tissues and organs.

Proteins: Biochemical Characteristics and Molecular Architecture

Biochemically speaking, elucidation of amino acid sequences for talin and vinculin was not particularly informative. In both proteins, the prevailing secondary structure elements are amphipathic α-helices which form series of helix bundles. There is no catalytic activity and, furthermore, intact proteins purified from tissues reveal no ligand interactions. Upon proteolytic cleavage or recombinant expression of putative domains, binding to F-actin, β-integrin and (other) cytoskeletal proteins can be observed, however, definition and functional characterization of protein-binding sites has proven complicated and is not completed to date. In recent years, structural analysis involving X-ray crystallography, NMR spectroscopy, and electron microscopy provided an increasing number of high-resolution datasets on domain folds and even ligand-bound domain structures. These structures helped understand the complexity of protein interactions and allowed the development of functional models. Molecular details of structure analysis performed on vinculin and talin are covered in recent reviews [4, 5].

Structure and Ligand-Binding Sites of Vinculin

For vinculin (116 kDa) a complete structure of its 'inactive' conformation is available [6]. In this conformation, the four helix-bundle domains (Vd1–Vd4) of the head (Vhead, ~90 kDa) form a pincer-like structure that covers large surface areas of helix-bundle Vd5, the so-called vinculin tail (Vtail, ~25 kDa). Vhead and Vtail are connected by a flexible proline-rich linker region (Vlinker). Ligand binding of both vinculin domains is blocked by the high-affinity head-to-tail interaction (HTI) with an estimated dissociation constant (K_d) of <1 nM [6, 7].

Binding of Vhead. Interaction of Vhead with other cytoskeletal adaptor proteins, like talin and α-actinin, requires conformational rearrangement of a 4-helix bundle in Vd1 to allow incorporation of one additional amphipathic helix provided by the binding partners. This process, termed 'helix bundle conversion', is inhibited while Vhead is bound to Vtail. Release of Vtail lifting sterical and conformational constraints is called 'activation' of vinculin [6, 8]. For Vd1, a second, alternative binding site was suggested by Izard et al. [9], which does not require bundle rearrangement and may be involved in the activation of vinculin. Other interaction sites of the vinculin head, e.g. binding sites for α- and β-catenin, are not characterized at the molecular level. The linker region Vlinker comprises several proline-based binding sites for proteins that affect actin dynamics like VASP and Arp2/3, and for other components of the cell adhesion complex. Availability of these binding sites is co-regulated by the HTI.

Binding of Vtail. In 'active' vinculin, Vtail being released from the head can provide a connection to the actin cytoskeleton. In an analysis combining electron microscopy and computational docking of crystal structures, Hanein et al. [10] observe that the helix-bundle Vd5 of Vtail contacts two neighboring actin monomers in the filament. Furthermore, F-actin binding involves dimerization of Vtail, which can contribute to F-actin bundling in adhesion sites [10, 11]. In vitro studies suggest that Vtail may also be involved in actin filament capping and thus influence regulation of actin turnover in cell contacts [12]. Other interaction partners of Vtail are paxillin, PKCα and acidic phospholipids, phosphatidylinositol-4,5-phosphate (PIP_2) and phosphatidylserine, of the plasma membrane. Competition of F-actin binding by acidic phospholipids, a feature of many actin-binding domains, may restrict the duration of vinculin's interaction with actin. The structure of Vtail has no similarities with other domain folds that bind to F-actin and/or acidic phospholipids [13, 14]. Therefore, an ongoing, EPR-based study, revealing different conformational rearrangements in Vtail induced by binding of both partners, may provide a better understanding of the mutual relationship of actin and lipid binding to vinculin [F. Dietrich, C. Abe, H.J. Steinhoff and W.H. Ziegler, unpubl. data]. In vitro, different parts of Vtail interact with acidic phospholipids. These are surface-exposed positively charged patches on Vtail, the 'basic ladder' and the 'basic collar', as well as the C-terminal hydrophobic arm, which can serve as docking site for PKCα in a lipid-dependent fashion [15–18]. The interaction of paxillin with Vtail although characterized biochemically to some extent [19] is not understood, and in many cell types difficult to detect. Furthermore, paxillin targeting to adhesion sites occurs independent of Vtail [20]. Together, biochemical and structural analyses of vinculin have defined a considerable number of binding sites that contribute to the function of the 'active' protein in the cell adhesion complex.

Structure and Ligand-Binding Sites of Talin

Talin (~270 kDa) consists of an N-terminal globular FERM domain, the talin head (~50 kDa), that is connected by a short linker to a large flexible C-terminal domain, the talin rod (>200 kDa). The rod domain comprises 62 amphipathic α-helices (H1–H62), which form a series of helix-bundle domains, and terminates in a dimerization site in helix 62 (H62). Ligand interactions of talin are inhibited by an intramolecular interaction of the FERM domain in the head with a helix-bundle domain of the rod [21, 22]. Although no complete model of talin exists, a combination of structural and biochemical analyses provided a quite detailed map of binding sites for diverse ligands, in particular two binding sites for β-integrin (IBS), three binding sites for actin filaments (ABS) and eleven α-helices allowing high-affinity interactions with vinculin (VBS) [5].

Integrin Binding. Of the two integrin-binding sites, IBS1 in the FERM domain, which is homologous to that in the band4.1/ezrin/radixin/moesin (FERM) family of cytoskeletal proteins, is well defined at the molecular level. The FERM domain of the talin head interacts with the membrane-proximal helix and the (proximal) NPXY motif of β-integrin cytodomains and contacts simultaneously acidic phospholipids of the plasma membrane [23, 24]. Together these interactions of the talin head critically contribute to the activation of integrins [25, 26]. In addition to β-integrins, other ligands, the hyaluronan receptor layilin and phosphatidylinositol-4-phosphate 5-kinase type 1 γ 90 (PIPK1γ90) [27, 28] can bind to the talin FERM domain in a competitive fashion and may contribute to the control of talin function. IBS2, the integrin-binding site of the talin rod, consists of two interconnected 5-helix bundles (helices H47–H56). β-Integrin binding of the IBS2 domain has been studied biochemically but not resolved structurally [29–31]. Since both IBS of talin interact with the same, membrane-proximal part of β-integrin cytodomains, they have to interfere, but implications of their functional interplay remain to be addressed.

F-Actin Binding. In vitro, different modes of talin interaction with actin filaments were observed but not resolved in terms of function [32]. Three ABS were localized biochemically, one each to the FERM domain of the head [33], the N-terminal third of the rod and the C-terminus of talin [34]. The ABS of the talin head is not sufficient to maintain the cytoskeletal connection [35, 36] and the second ABS is not elucidated structurally. Actin binding of the C-terminal site, in a conserved fold called THATCH domain (or I/LWEQ module), has a number of interesting biochemical features [37, 38]. The C-terminal ABS consists of a 5-helix bundle (in talin H57-H61) and a dimerization helix (H62). The first helix H57 stabilizes the bundle in a low-affinity state. F-actin is bound through a conserved hydrophobic surface of helices three and four (H59, H60). The last helix (H62) is also involved in the actin interaction, and (antiparallel) dimerization of this helix is required for actin binding of the domain [39, 40]. The talin THATCH domain associates with three actin monomers of the same filament. Thus it does not support F-actin bundling or cross-linking. Furthermore, the C-terminal ABS is pH-sensitive and may allow regulation of actin binding via Na^+/H^+ antiporter activity in cell contacts [41].

Vinculin Binding. Detection of binding sites for vinculin has proven difficult. Initially, three amphipathic α-helices of the talin rod (VBS1-VBS3) with Vhead binding affinity were described [42]. Co-crystals of VBS helices with Vhead revealed that binding of vinculin depends on the hydrophobic pattern on one surface of the amphipathic helix [8, 43], leading to high-affinity interactions with K_d ranging from ~20 to 80 nM. Analysis of Vhead binding to all 62 amphipathic helices, using an array of spot-synthesized 25mer peptides, resulted in 11 high-affinity VBS helices, each of which being a potential binding site for vinculin [44]. To allow Vhead binding, however, structural rearrangement of the helix-bundle domains in the talin rod are required and bundle stability renders isolated, intact domain bundles mostly inaccessible to Vhead [45, 46]. In cell contacts, different mechanisms including protein phosphorylation, membrane interaction and/or mechanical stress may allow binding site activation. Different approaches involving molecular dynamics simulation [47, 48] as well as optical tweezer-based stretching of talin rod constructs [49] suggest that mechanical force can release VBS helices, making Vhead binding a potential mechanism of stress-sensing in adhesion sites. Consistently, vinculin binding to adhesions sites increases upon application of mechanical stress [50, 51]. Whether or not vinculin-induced rearrangement of helix bundles in the talin rod is reversible, and how the other helices that are not bound to Vhead, are stabilized or reorganized, remains to be resolved [45, 52].

Over the last two decades, identification of binding sites in talin has proved difficult. In particular, the domain structure of the rod is still not fully established. In addition, many ligand-binding sites are concealed or in a low-affinity state in 'generic' helix-bundle constructs of the rod. Hence, description of talin function(s) cannot be reduced to a sequence of domains folds with known, defined functionality.

Cell Adhesion Complex: Protein Properties in the Cellular Context

Biochemical and structural analyses of vinculin and talin, although incomplete in some details, reveal a complex architecture and regulation of ligand-binding sites of both adaptor proteins in our simple model of the cell adhesion complex. The functional relevance of specific protein properties can be determined only in the cellular context. In particular the rigorous control of high-affinity binding sites, which leads to almost 'inert' cytoplasmic conformations, appears to be of critical biological relevance, and binding sites observed in vitro should be treated as options that require verification and functional interpretation in the context of cell adhesions.

Approaches addressing the relevance of protein interactions frequently involve fluorescently tagged protein variants that allow (i) cellular localization of the protein, as well as (ii) time-resolved analysis of interactions and/or (iii) determination of conformational activation [53]. GFP-tagged constructs of cytoskeletal proteins were widely used in fluorescence recovery after photobleaching (FRAP) analysis, which allows determination of half-lives (or residency times) of protein variants in cell contacts [54, 55]. In addition, some constructs using Förster resonance energy transfer (FRET) were developed for vinculin and other adhesion complex proteins like FAK to monitor conformational rearrangements that separate or reorientate protein domains [56, 57]. In studies using these semiquantitative techniques of fluorescence microscopy, protein constructs are frequently investigated in mature integrin-based cell extracellular matrix contacts – so-called focal adhesions (FAs). In these adhesions, which live typically 20–30 min in resting cells, a steady-state equilibrium of protein in the cell adhesion complex is assumed. FRAP analysis of exchange kinetics in FAs indicates that different pools of vinculin and talin exist in adhesion sites, which were suggested to represent tethered and bound states of protein [58]. Furthermore, lifetime and conformational analyses require that exogenous protein variants become incorporated in the steady state of the cell adhesion complex, a condition which is difficult of control. Morphology, frequency and distribution of adhesion sites are used as cues to assess effects of (exogenous) protein expression on cell adhesion.

Activation of Adaptor Proteins in Cell Adhesions

Conformational activation of vinculin in the cell adhesion complex was predicted for a long time. However, the design of a vinculin-conformation sensor was technically and biochemically very challenging. Sue Craig and her laboratory [59] successfully established a FRET-based sensor construct using an internal YFP, positioned in the strap of Vtail, and a CFP at the C-terminus of vinculin. This CFP/YFP-tagged vinculin variant was purified and characterized biochemically to document that (i) F-actin binding of Vtail required coactivation of a Vhead ligand and, furthermore, that (ii) reduction of FRET efficiency, signaling vinculin activation and F-actin binding, was observed only, when head and tail ligands interacted simultaneously with the sensor. Upon expression of the conformation sensor in cells, FRET signals revealed 'active', F-actin-bound vinculin mostly restricted to adhesion sites. The sensor was used to report F-actin-bound vinculin in cell contacts of resting and spreading vinculin-null fibroblasts, which are deficient for endogenous vinculin, as well as in adhesion sites of slowly moving smooth muscle cells, which due to a lower sensitivity to phototoxicity allowed observation of the sensor over prolonged periods of time [59]. Importantly, consistent with the concept of different conformational options and binding partners, vinculin was not detected exclusively in its actin-bound conformation in adhesion sites. Other conformation(s), which are not assessed by the sensor, may relate to lipid-bound Vtail or a 'closed', tethered state of vinculin, making the protein readily available for turnover. Further hints for functional diversity of vinculin in the cell adhesion complex came from observation of dynamic contacts in smooth muscle cells. In gliding or disassembling FAs, actin-bound vinculin localized preferentially to the proximal edge where maximal force is expected [59]. The different options of vinculin for interactions in adhesion sites clearly request further investigation, to elucidate properties related to specific protein function(s). Independent confirmation of vinculin activation in adhesion sites was established using vinexin-β recruitment. Upon ligand interaction of Vhead, vinexin-β can bind to the proline-rich region of Vlinker but not to the inactive conformation of vinculin. In a model using permeabilized vinculin-null cells, vinexin-β was localized in adhesion sites decorated with exogenously supplied, full-length vinculin but not to those decorated with Vhead lacking the proline-rich region. Thus, vinexin-β targeting correlates with active vinculin in adhesion sites [59].

The high-affinity HTI of vinculin provides a well-established example of conformational protein regulation conveyed by cell contact association. Structural, biochemical and biophysical data of talin indicate that in a similar way the head-to-rod interaction and cellular functions of talin rod domains may be regulated through coordinated ligand interaction in the cell adhesion complex. Elucidation of these processes awaits development and application of suitable 'protein tools'.

Consequences of Constitutive Vhead Binding

The vinculin HTI depends on two cooperative interactions of Vhead domains Vd1 and Vd4 with Vtail (Vd5). Together these interactions result in an extremely low equilibrium K_d estimated at <1 nM [6, 7]. Since K_d values for binding of talin (and talin rod) were determined experimentally in the range of 100 nM, interaction of Vhead with talin-VBS will be readily displaced by Vtail, unless Vtail is interacting simultaneously with a different binding partner like actin. The requirement of vinculin HTI for combinatorial input of several ligands was demonstrated in a number of biochemical approaches and was shown to prevent efficiently vinculin activation outside of adhesion sites [60, 61]. Conversely, vinculin variants with 100-fold reduced HTI (K_d ~100 nM) perform enhanced interactions revealed by a 2-fold increase of half-life times in adhesion sites, as compared to wild-type vinculin. This is consistent with a shift of the equilibrium towards constitutive VBS-binding of Vhead. Resting cells expressing constitutive vinculin HTI mutants display altered adhesion site distribution and morphology. FAs cover the entire ventral surface of these cells, and in the cell body, no diffuse fluorescence intensity of the GFP-tagged constructs is observed, reflecting strong adhesion sites targeting of the mutants. Furthermore, vinculin HTI mutants (and Vhead constructs) can control the residency times of talin and integrin in FAs, but not the half-lives of other proteins like α-actinin and paxillin [20, 60]. The enhanced half-life of integrin/talin/vinculin HTI mutant complexes leads to FA stabilization and reduced turnover. Contact stabilization is accompanied by a 2- to 3-fold increase in the number and size of FAs, and in the total adhesion area of cells. Importantly, the impact of vinculin HTI mutants on adhesion sites depends on Vhead binding to talin. The Ala50-to-Ile (A50I) mutation, which reduces the affinity of Vhead for talin VBS helices, abrogates the effects of HTI mutants [20]. Furthermore, uncontrolled Vhead activity can tip the steady state of protein interactions towards constitutive talin/integrin interaction and uncouple cell contact dynamics from cytoskeletal control. Constitutive binding of Vhead induces and stabilizes both talin activation and integrin clustering [20]. This stabilization of cell contacts is uncoupled from mechanical stress and the cytoskeletal connection, since Vhead-induced contacts are resistant to pharmacological inhibition of actomyosin contractility and disruption of actin filaments [20]. Thus, vinculin HTI imposing on vinculin the requirement for a combinatorial input from different ligands provides one of the control mechanisms of the cell adhesion complex that balance the adhesiveness of cells.

Functional Modules: Cooperation of Binding Sites

In talin, head-to-rod interaction is assumed to regulate the distribution between the free, cytoplasmic and the bound, localized conformations in a global fashion. However, further levels of activity regulation may be required. Analysis of binding site activity in the talin rod provides evidence that adhesion site targeting is efficiently mediated by a module of cooperating domains situated at the C-terminus of talin. This module comprising the IBS-2 domain (helices H47–H56) and the C-terminal ABS/dimerization domain (H57–H62) is termed talinC. Evidently, a number of binding sites in talin can lead to adhesion site localization, but once activated, independent and firm interaction of multiple binding sites would lock the protein in the adhesion complex and prevent dynamic exchange with the cytoplasm. This is not observed. FRAP-based analyses of talin indicate that ~80% of protein localized in FAs exchange with a half-life <1 min and thus far below the turnover time of the adhesion. Hence, it appears that modules allowing regulation of binding sites in a coordinated fashion render the protein amenable to (cellular) control.

In talin-deficient cells, expression of the talin head does not recover defects in adhesion site formation and force coupling of the cytoskeleton [36]. This is consistent with results from a transgenic model in *Drosophila*. Studying muscle attachment sites of *Drosophila* embryos, Tanentzapf and Brown [35] demonstrate that a talin mutant with defective β-integrin binding of the head, as expected, reduces the strength of integrin-mediated adhesion to extracellular matrix. However, talin recruitment by integrins and its ability to connect integrins to the cytoskeleton, both being also functions of the talin rod, are not disrupted. Accordingly, localization experiments based on talin domain constructs reveal that the talin head localizes rather weakly, while some constructs of the talin rod in particular talinC (H47–H62) target very effectively to adhesion sites [58]. Combining biochemical characterization of talin domains and FRAP-based analysis of exchange kinetics in adhesion sites, my laboratory observed that low-affinity interactions of talin domains with β-integrin and F-actin convey adhesion site targeting of talinC. In the cell adhesion complex, concomitant engagement of the IBS-2 and the C-terminal ABS domain leads to conformational activation and high-affinity binding of both domains. This activation is reversible, and half-life and adhesion sites dynamics of talinC are similar to those of full-length talin [58].

Interaction of vinculin with talinC, which contains two high-affinity VBS helices H50 and H58, is not fully understood. Results collected in different cell types including vinculin-null fibroblasts suggest that vinculin can affect both talinC activation, supporting conformational rearrangement, and talinC inactivation, competing helix H50 interaction of β-integrins in the IBS-2 domain [58, 62]. Furthermore, we observed that control of F-actin binding to the THATCH domain is critical for talinC function. When the C-terminal ABS domain (H57–H62) is expressed without the inhibitory helix H57 (also called 'upstream helix'), F-actin interaction of the truncated domain construct shifts the steady state of the cell adhesion complex thereby affecting severely the morphology and the distribution of cell adhesions as well as the organization of the actin cytoskeleton [58]. In talinC, control of the cytoskeletal connection is linked to integrin binding of the IBS-2 domain. Engaged IBS-2 is suggested to modulate helix H57 interaction within the C-terminal ABS domain (H57–H62) and thus to allow reversible, high-affinity F-actin binding. Conversely, biochemical data and FRAP analysis indicate also that the strength of integrin binding depends on the ABS domain. High-affinity integrin binding induces rearrangements in the IBS-2 domain (H47–H56), whereupon vinculin gains access to the VBS helix H50, and competing the integrin interaction, vinculin can initiate inactivation and release of talinC [58].

The talinC module serves a paradigm showing that a combination of binding sites in their low-affinity state can effectively restrict interaction and high-affinity binding to the cell adhesion complex. Mutual dependence of binding activities in the module guarantees 'self-regulated' inactivation. In talinC, competition of actin binding through acidic phospholipids or a rise in pH, in addition to vinculin competition of integrin binding, ensure 'dynamic' instability of interactions. Ligand interactions coordinated in the talinC module thus allow induction of high-affinity binding sites in both the IBS-2 and the C-terminal ABS domain in a controlled and balanced fashion, which does not jeopardize the equilibrium of protein turnover in the cell adhesion complex.

Different aspects of talin function such as targeting to adhesion site, activation of integrins, and control of scaffold properties as well as the cytoskeletal connection may require independent, differential regulation and organization of the protein in a number of functional modules. Although open questions remain, the talinC module is one example for which coordination of binding site activities was established. The second is the FERM domain in the talin head. This domain controls integrin activation and sequesters other transmembrane proteins. Cellular control of the talin head will be discussed in more detail later. Other domains of talin are expected to be organized as well in modules allowing control of high-affinity interactions and stepwise (modular) regulation of talin activity.

Scaffold Function and Signaling

Analyses of interactions addressing scaffold function of talin or vinculin in the cell adhesion complex are limited. The vinculin-paxillin interaction is one of the best studied examples. Although insufficiently characterized at the molecular level (as discussed above), experiments in vinculin-deficient cells indicate that the interaction of vinculin with paxillin is linked functionally to adhesion site signaling. Furthermore, the interaction was suggested to modulate ERK signaling and adhesion site dynamics, thereby affecting cell motility and generation of traction forces [63, 64]. Consequences of vinculin deficiency on the properties of the cell adhesion complex were analyzed in parental F9 embryonal carcinoma cells as well as in F9 cells deficient for vinculin (F9 vinculin null cells) [65, 66]. To address the relevance of specific vinculin interactions, cells were reconstituted with full-length protein and domain constructs, Vhead (aa 1–821) or Vlinker+Vtail (aa 811–1066). Compared to parental cells, F9 vinculin null cells are characterized by small short-lived FAs, a defect in cell spreading, and 2-fold enhanced cell migration [66, 67]. In addition, cells display increased activity of the extracellular signal-regulated kinase (ERK) pathway, which seems to play a central role in FA regulation of vinculin-deficient cells [63, 68]. ERK was shown to phosphorylate and activate myosin light chain kinase [69] and calpain II [70], which leads to adhesion site dissolution and enhanced cell migration [71, 72]. Reconstitution of vinculin null cells with vinculin or the Vlinker+Vtail construct restores wild-type levels of ERK activity and cell motility.

Subauste et al. [63] suggested that Vtail competition of the FAK-paxillin interaction is required to restrict ERK activity. Upon adhesion of cells to extracellular matrix, FAK and paxillin are attracted to nascent cell contacts where they interact. Enhanced tyrosine phosphorylation of both proteins triggers signals that stimulate ERK activity and adhesion site turnover as verified by site-directed mutagenesis of paxillin and pharmacological inhibition of the ERK pathway [63, 68]. Pull-down experiments in F9 vinculin null cells analyzing protein-protein binding indicate that the constitutive FAK-paxillin interaction of adherent cells is reduced in the presence of vinculin and the Vlinkler+Vtail construct, but not when Vhead is expressed [63]. Thus, active vinculin in FAs appears to modulate protein-protein interactions and signaling pathways, which control turnover of the cell adhesion complex.

Biophysical analyses in the F9 cell model, employing fibronectin-coated paramagnetic beads and magnetic tweezers to study mechanical properties of cells, support this view. Mierke et al. [64] reported that Vlinker+Vtail expression in F9 vinculin null cells not only restored spreading and stiffness of cells, but importantly also elevated tension generation and traction forces, both of which were strongly reduced in vinculin-deficient cells. In contrast, Vhead which can stabilize integrin-talin complexes in the active conformation, did not rescue traction forces. Biophysical analyses testing force transmission across the cell adhesions complex indicate that vinculin is involved in but not essential to mechano-coupling of cell adhesions. However, by fine tuning the paxillin-FAK interaction, it may have a pronounced impact on the stability of FA, the actomyosin contractility and the generation of traction forces required for tissue invasion of cells [73, 74]. Thus, it appears that vinculin interaction in the cell adhesion complex is not only relevant to maximal force transmission but also to the balance of force generation and motility, adjusting adhesive properties of FA.

Dynamic organization of the cell adhesion complex even for selected well-studied proteins is far from being understood. Recently, Wolfenson et al. [92] using a combination of FRAP studies and mathematical modeling suggested four dynamic states for vinculin and paxillin. These include an immobile FA-bound fraction, an FA-associated fraction undergoing exchange, a juxtamembrane fraction experiencing attenuated diffusion, and a fast diffusing cytoplasmic pool. The authors propose that the space adjacent to cell adhesions, which they termed juxtamembrane domain, is involved in FA regulation. How the different protein states affect each other and contribute to the self-organization of the cell adhesion complex, and how modules of protein domains coordinate binding site activities and determine function, are attractive and challenging questions of ongoing research.

Cell Migration: Regulation of Adaptor Proteins during Contact Turnover

In this discussion of adaptor protein properties and functional regulation, so far, mostly cell adhesion complexes in steady state were considered. In motile cells, however, turnover of cell adhesions is essential and integrin-based contacts vary considerably in protein composition and function during their life cycle. Initial contact formation and maturation of adhesion sites in protrusive areas of motile cells require different functional interaction of adaptor proteins than stably engaged or dissolving cell contacts. Signals that were shown to adjust protein interaction are (i) mechanical coupling and actomyosin contractility, (ii) protein phosphorylation, (iii) generation and release of phosphoinositides (or more generally speaking acidic phospholipids), and (iv) proteolysis. These signals will be discussed in relation to talin and vinculin, and their effects on function and turnover of adhesion sites.

In different cell types and tissues, specific requirements are expected to generate a huge variety of different adhesion complexes with unique protein composition and functionality, however some of the core properties should be universal. To outline key principles of adhesion site regulation, Geiger et al. [75, 76] suggested and characterized three main states of integrin-based cell-extracellular matrix adhesions with distinct (adaptor) protein composition, dynamics and function. Focal complexes (FXs) are transient, dot-like adhesions, which form in a hierarchical manner under the outward moving lamellipodium. FXs turn over and disappear within less than a minute, unless force coupling supports their transition into FAs. These are considerably larger structures that are connected to actin- and myosin-containing stress fibers. Actomyosin contractility is essential for the persistence of FAs, as well as their transformation into fibrillar adhesions (FBs). These are elongated structures characterized by prominent components, the fibronectin receptor (integrin $\alpha_5\beta_1$) and tensin. FBs perform a sliding centripetal movement towards the cell center and play an active role in fibronectin fibrillogenesis.

Formation of the three different integrin-based adhesions, FX, FA and FB, is modulated by members of the Rho GTPase family, in particular the balance between Rac and Rho activities [77, 78]. Active Rac stimulates actin dynamics in the lamellipodium and is required for FX formation, whereas regulation of actomyosin contractility by Rho induces FA formation and growth, and supports development of FBs. Tyrosine phosphorylation is a characteristic of all three types of adhesions, even though in FXs its functional relevance is not resolved. The interaction between Src family kinases (SFK), which are responsible for initial tyrosine phosphorylation of FXs, and FAK is of central importance for the control of adhesion site maturation and turnover. FAK is attracted during FA formation and its activity is intimately linked to the transformation process leading to mature FAs, which are mechanically connected to the cytoskeleton [50, 79, 80]. The different properties of these adhesions and their transformation into another, although not fully understood at the molecular level, evidently require dynamic control of adaptor proteins activities.

Observation of Adaptor Proteins in Dynamic Adhesions Sites

Using quantitative fluorescence microscopy, Zaidel-Bar et al. [76] addressed the protein composition of nascent adhesion sites, evaluating a number of well-characterized components of cell adhesion complexes. During initial contact formation at the leading edge of migrating cells, they observe a hierarchy of events related to the establishment of FXs and their transition into FAs. Relative fluorescence intensity profiles for proteins in both structures indicate that FXs are initiated by integrin binding to extracellular matrix followed by attachment of talin and paxillin. Transition into FAs is associated with a rise in paxillin, vinculin, and FAK as well as further adaptor proteins. In particular, zyxin was identified as a marker of mature FAs, for example in retracting areas where adhesion sites experience mechanical stress [76]. Consistently, zyxin localization and kinetics in FAs correlate directly with force coupling of cell contacts, whereas vinculin that accumulates during FA assembly is less acutely regulated [81].

'Spatio-temporal image correlation spectroscopy' (STICS) [82] and 'fluorescence speckle microscopy' in combination with 'total internal reflection microscopy' (TIR-FSM) [83], two advanced methods of image analysis and protein tracking, provided further insight of how talin and vinculin contribute to the coupling of the cytoskeleton. At the leading edge of moving cells, the F-actin network of the lamellipodium is characterized by a fast retrograde flow of actin towards the cell center. In contrast, the flow of actin in the lamella just behind the lamellipodium is slower and the lamellipodium-lamella junction coincides with the appearance of FAs. Hence, cytoskeletal adaptor proteins providing docking sites connected to the extracellular matrix are expected to reduce actin dynamics. To compare direction and speed of the actin flow with the movement of integrins and adaptor proteins, Waterman-Storer et al. [83] employed TIR-FSM. The authors observe differences in the behavior of adaptors that bind actin as compared to the other components of the cell adhesion complex. Whereas movement of the latter, e.g. FAK, zyxin, and paxillin, is incoherent, relatively slow and similar to that of integrins, F-actin-binding proteins, e.g. α-actinin, vinculin and talin, reveal differential correlation of their motion with that of the actin network. α-Actinin, an F-actin bundling protein that localizes also in adhesion sites, directly follows the actin flow, whereas vinculin and talin display intermediate motion reflecting dynamic association of these proteins with F-actin and the cell adhesion complex. Vinculin moves rather slowly with good correlation to the directionality of the actin flow, while talin moves fast but less coherent and correlated [83]. Talin's behavior may relate to transient receptor interaction and the flexible conformation of talin dimers. Observation of vinculin and talin dynamics suggested that they provide a slippage interface or molecular clutch which connects highly dynamic F-actin structures to relatively static adhesion patches [83]. STICS analysis confirms the intermediate movement of both adaptor proteins in relation to fast-moving actin and α-actinin, and the slower core adhesion complex, as well as the good correlation of vinculin movement with the actin flow [82]. Furthermore, kymographs addressing the movement of vinculin domain constructs at the leading edge of cells also support the interface concept. An HTI-mutant of vinculin, vinculinT12, and Vtail were observed to perform targeted movement, consistent with their connection to the flow of the actin network, whereas Vhead lacking the F-actin-binding site of Vd5 (Vt) stayed in position [20]. Evidently, the slippage interface reflects the regulation of ligand-binding activities in talin and vinculin.

Force and Phosphorylation: Effects on Adaptors and FA Regulation

The critical relevance of talin for the coupling of integrins to the cytoskeleton was established in optical tweezer experiments. In the optical trap, biophysical properties of nascent adhesion sites can be studied with the help of extracellular matrix-coated beads, which allow application of defined forces to the integrin connection. After binding of talin, early events in focal complex formation and force coupling involve recruitment of vinculin (and paxillin), as well as activation of tyrosine kinases, SFK and FAK [50, 79]. In talin-deficient cells, activation of tyrosine kinases was observed upon integrin binding to extracellular ligands, indicating that kinases can be activated independent of talin engagement and force coupling [79]. Conversely, to allow strengthening of the initial integrin-cytoskeleton linkage, control of FAK activity by the phosphotyrosine phosphatase Shp2 was required and shown to reduce exchange rates of adaptor proteins [84].

For talin, the relationship between tyrosine phosphorylation and mechanical coupling in cell adhesions is not understood at the molecular level, and further elucidation suffers from two obstacles. Firstly, although phosphorylation sites in talin have been discussed for a long time, and in vivo sites have been mapped in talin protein isolated from activated platelets [85], the functional relevance of specific phosphorylation sites remains elusive. Secondly, only very recently, cells truly deficient for talin have become available. There are two genes for talin, *TLN1* and *TLN2*, encoding homologous proteins with very similar domain structure, talin1 and talin2. Talin1 is essential for integrin-mediated cell adhesion and predominant in many tissues and cell lines, whereas the role of talin2 is unclear [86]. However, in talin1-knockout cells, talin2 expression is activated and partially recovers adhesion and spreading defects. Therefore, talin2 expression needs to be controlled by siRNA knockdown, to fully suppress talin function in talin1-knockout cells. Using this approach, Zhang et al. [36] recently reanalyzed the process of cell spreading in cells deficient for talin1/2, and in cells reconstituted with talin (talin1) or the talin head. The authors report that early spreading of cells on fibronectin depends on integrin binding only, which is accompanied by the activation of SFK-linked signaling and actin polymerization. Thereafter, actomyosin contractility is activated and full-length talin is required at this point for the reduction of the actin rearward flow and coupling of peripheral actin to engaged integrins. Activation of FAK and myosin II-dependent reorganization of the peripheral actin network accompany the assembly of FAs. Mechanical coupling of the talin rod (in full-length talin) to the cytoskeleton is essential in this process. When talin1/2-deficient cells are reconstituted with the talin head only, the FERM domain can stabilize (and activate) integrin binding to the extracellular matrix, however, only weak forces are generated, mature FAs do not form and eventually, cells round up due to actomyosin contraction [36]. Interestingly, although FAK and the talin head can directly interact [87], expression of the talin head does not rescue Y397 phosphorylation and activation of FAK. Y397 phosphorylation depends on myosin II contraction-mediated actin reorganization, indicating that traction force is involved in the activation of FAK [36]. Furthermore, FAK activation appears to provide a positive feedback loop. Garcia et al. [88] reported that FAK activity can enhance integrin activation and the rate of adhesion site strengthening in a fashion dependent on talin. Noteworthy, expression of FAK does not alter the rate of vinculin incorporation (accumulation) during FA assembly, however the maximal vinculin level is reduced by one third in FAK expressing as compared to FAK-deficient cells, indicating a possible effect of phosphorylation on vinculin exchange kinetics.

Recently, effects of phosphorylation in vinculin were addressed in adhesion sites and elucidated to some extent. Similar to talin, phosphorylation of vinculin purified from cells is extremely low and despite many efforts has proven difficult to study. It is assumed that phosphorylation occurs very transiently in cells and is tightly controlled by protein phosphatases. Using vanadate treatment to block phosphatase activities, Haimovich et al. [89] detected SFK-dependent phosphorylation of tyrosines Y100 and Y1065 in vinculin. Biochemical analysis of purified protein further indicated that phosphorylation of the C-terminal Y1065 can influence the affinity of the vinculin HTI, but does not induce F-actin binding of the autoinhibited full-length protein. Moreover, in the inactive conformation of vinculin, Y1065 is buried by Vlinker and prevents Src phosphorylation in vitro. Thus, Y1065 phosphorylation appears to require activation of vinculin in adhesion sites, which may involve priming phosphorylation of adjacent sites, serines S1033 or S1045, by protein kinase C [17, 89]. It was speculated that transient phosphorylation of vinculin can influence its interaction with other proteins in the cell adhesion complex and modulate the lifetime of the open, active conformation [89]. Consistently, analysis of vinculin phosphorylation in migrating keratinocytes using the vinculin Y1065 to Phe mutant (vinculin-Y1065F) revealed effects of tyrosine phosphorylation on cell motility and force transmission. Using FRAP analysis, Hoffmann et al. [90] show that residency time and recovery of vinculin are different in short-lived FAs at the leading edge of migrating cells as compared to larger (mature) FAs in the rear or in sessile cells. Enhanced turnover of adhesion sites at the leading edge correlates with reduced residency times of vinculin and a smaller immobile fraction of the protein. When vinculin-Y1065F was expressed and analyzed at the leading edge, the phosphorylation-defective mutant displayed reduced exchange dynamics similar to those observed for wild-type vinculin in sessile cells. Thus, Y1065 phosphorylation appears to enhance exchange rates of vinculin in short-lived FAs. In addition, comparison of vinculin-Y1065 phosphorylation levels revealed that phosphorylation is high in motile cells and barely detectable in FAs of stationary cells. In migrating cells, the phospho-vinculin to vinculin ratio is highest in FAs close to the leading edge and decreases uniformly towards the rear. Furthermore, force generation determined from substrate deformation is inversely correlated. In motile keratinocytes, strong forces are concentrated at the rear, which is characterized by low phospho-vinculin levels and stable FAs [90].

Analysis of tyrosine phosphorylation in paxillin, positions Y31 and Y118, provided similar results. Using phosphomimetic (Glu, E) and phosphorylation-defective (Phe, F) mutants paxillin-2E and paxillin-2F, Zaidel-Bar et al. [91] observed that paxillin-2E stimulates assembly and turnover of FA, and favors FAK binding, whereas dephosphorylated paxillin or paxillin-2F is required to allow stabilization of adhesion sites and transition of FAs into FBs. The authors suggest a model wherein phosphorylation of paxillin is required during formation and turnover of adhesion sites, whereas mechanical forces are needed to suppress FA turnover and protein dynamics leading to stabilized adhesion sites. This view is consistent with the analysis of force coupling by talin [84] and may constitute a general pattern for the regulation of adaptor proteins by SFK/FAK signaling. Though questions remain at the level of protein-protein interaction, controlled phosphorylation of adaptor proteins is clearly essential for FA dynamics and force coupling.

As discussed previously, results from Subauste et al. [63] indicate that vinculin HTI and release of Vtail are involved in the control of kinase activity and paxillin

FAK phosphorylation. This highlights vinculin's impact on adhesion site dynamics and may shed a different light on results reported by Garcia et al. [51]. Adhesion analysis of starved cells on micropatterned surfaces (providing a defined adhesion area) revealed that induction of actomyosin contractility by serum stimulation was associated with a huge 4-fold increase of vinculin and doubling of talin protein in adhesion sites, which is consistent with the well-defined function of the vinculin-talin interaction in FA organization and strengthening. However, the strength of adhesion rose by 30% only [51]. This moderate impact becomes much less surprising when vinculin functions other than force coupling are considered. Consistently, differential interaction of vinculin was revealed by the heterogeneous signal of the vinculin FRET sensor in FAs [59]. In addition, FRAP analyses not only demonstrated different exchange rates of vinculin between the front and the rear of migrating cells [90] but also between the proximal and the distal ends of large FAs [92], reflecting variable input of force, phosphorylation and binding partners.

Recruitment and Release of Adaptors: Binding Partners, Lipids and Proteolysis
Being critically involved in integrin receptor activation and adhesion site strengthening, talin obviously requires a tight regulation of its ligand-binding activity. But, molecular interactions that contribute to the inactive, cytoplasmic conformation of talin are only partially understood. Recently, an interaction of the FERM F3 subdomain in the talin head with the talin rod was mapped and characterized. X-ray crystallography and NMR spectroscopy of domain interactions provided an explanation for the intramolecular inhibition of integrin binding to the talin head (IBS1). The talin F3 domain binds to a 5-helix bundle of the talin rod (helices H37–H41) in a fashion that (i) partially masks the interaction site for β-integrins and (ii) sterically inhibits interaction of the FERM domain with the membrane, both of which inhibit talin FERM binding to integrins. In contrast, the talin F3 interaction site for PIPK1γ90 remains exposed and may provide a route to talin activation [21, 22]. In the rod, the talin F3 interaction site is adjacent to the VBS3 5-helix bundle (H42–H46) and the IBS2 of talinC (H47–H62). Whether the talin F3/rod interaction is the only head-to-rod interaction, and how ligand-binding activities of the rod are controlled, remains to be resolved. Despite these open questions, a mutant defective in talin F3/rod binding, talin M319A, revealed that such interaction indeed regulates talin activity. In CHO cells expressing integrin $\alpha_{IIb}\beta_3$, the M319A-mutant of full-length talin was almost twice as effective at activating the integrin as wild-type talin [21].

Activation of talin requires recruitment to the (plasma) membrane and interaction of talin with PIP_2, which completes the talin F3/rod interaction [21, 93]. In addition, the talin activation process is tightly linked to the control of integrin activity and most likely involves a number of different ligand partners and signaling events. Analysis of integrin $\alpha_{IIb}\beta_3$ activity in platelets established an important role of the small GTPase Rap1A [94]. Together with the adaptor protein RIAM, Rap1A promotes recruitment of talin and activation of $\alpha_{IIb}\beta_3$ integrins in a PKCα-dependent fashion [95, 96]. Other

RIAM-related proteins like lamellipodin are less understood and may provide Rap1A-independent targeting of talin [97]. Alternatively, or in addition since a relation to Rap1A/RIAM has not been established, PIPK1γ90-dependent events promote talin targeting and activation. Talin binds to and activates PIPK1γ90 [98, 99]. Furthermore, FAK and Src signaling enhance the interaction of both proteins and their translocation to the plasma membrane [100]. There, PIP_2 produced by PIPK1γ90 may facilitate the release of the talin F3/rod interaction and initiates β-integrin binding [21]. Consistently, PIPK1γ90 activity was shown to promote FA assembly during cell spreading [100]. The major problem of this model consists in the affinity of talin F3 for PIPK1γ90, which is considerably higher than that for β-integrin cytodomains [27, 28], and would demand a mechanism that allows displacement of PIPK1γ90 from talin. Thus, the mechanism of talin activation requires further analysis.

In migrating cells, control of integrin activity also requires efficient inactivation signals that balance adhesiveness. Cleavage of adaptor proteins in the cell adhesion complex by calpain and subsequent proteasomal degradation is one of the better characterized signals that reduce cell adhesion at the leading edge and in the rear of cells. EGF stimulation of cell migration involves ERK-dependent phosphorylation and activation of calpain II [70]. Furthermore, calpain-mediated cleavage of talin was shown to be essential for FA turnover and cell migration [101] and appears to adjust FA dynamics in a fashion dependent on the talin head [102]. Calpain II cleavage in the neck region of talin liberates the talin head domain, which can subsequently localize to FA, activate integrins and maintain cell edge protrusions [36]. Clearly the activity of the talin head needs to be contained in cells, and Ginsberg et al. [102] recently provided the first evidence that cellular levels of the liberated talin head are modulated by antagonistic action of the E3 ubiquitin ligase Smurf1, and Cdk5, a cyclin-dependent protein kinase. Smurf1 binding to the talin head leads to ubiquitylation and proteolysis. When the talin head becomes phosphorylated at Ser 425 by Cdk5, Smurf1 binding is suppressed and the head domain is protected from degradation. Analysis of talin head phosphorylation mutants S425A and S425D in migrating cells reveals that the Smurf1/Cdk5 axis modulates a delicate balance of head stabilization versus degradation. If the equilibrium is tipped too far in either direction, FA dynamics change and cell migration is reduced.

Thus, the talin head was characterized as a module that allows functional control of adhesion site dynamics independent of the rod and the cytoskeletal connection. How the talin rod is processed after calpain cleavage, and whether this also generates protein modules of independent functionality, remains to be analyzed. Moreover, calpain cleavage can selectively affect cytoskeletal adaptor proteins and modulate their function [103]. Proteolytic cleavage evidently requires control of the liberated domains and several lines of evidence suggest that vinculin is most likely regulated by different means. In vitro, vinculin is cleaved in Vlinker by a number of proteases, but proteolytic Vhead products are rarely detected in cell extracts. This may reflect immediate and effective degradation of vinculin domains, but in vitro degradation experiments indicate that Vhead becomes resistant to proteolysis when bound to talin VBS helices [45, 58]. Consistently, Vhead domain constructs, even when expressed at moderate to low levels, localize in cell adhesions and suppress FA dynamics [20], corroborating that degradation of liberated Vhead cannot be very effective. Nevertheless, irreversible processes like platelet aggregation or cell damage may provide exceptions, and calpain-dependent cleavage of vinculin was observed in activated platelets and a cell damage model for pancreatic cells [104, 105].

The mechanism of vinculin recruitment is not fully understood at the level of molecular interactions. Similar to talin, vinculin may require specific partners to facilitate adhesion site localization, but these have not been identified yet. In inactive, cytoplasmic vinculin, the binding site(s) for VBS helices in Vhead and for F-actin in Vtail are inhibited, but low-affinity interactions to actin [10] and to VBS helices, via the alternative site of Vhead [9], are possible and may provide routes to vinculin activation. Adhesion site dynamics of vinculin are clearly dominated by the vinculin HTI, and Y1065 phosphorylation of vinculin, which was proposed to regulate the interaction, enhances protein exchange rates as required for increased FA turnover [60, 90]. On the other hand, Vhead forming a stabilized complex with talin can inhibit FA turnover in a dominant fashion [20]. Therefore, a contribution of vinculin cleavage to cell motility is not very likely. Instead, a balanced combinatorial input of Vhead and Vtail ligands is expected to adjust vinculin activity in the cell adhesion complex [6].

In this model of vinculin regulation, Vtail interaction with PIP_2 was proposed to initiate inactivation of vinculin. Competition of F-actin binding by locally generated acidic phospholipid (in particular PIP_2) may critically control (i) vinculin ligand binding, (ii) coupling of the actin cytoskeleton, and (iii) turnover of adhesion sites. As discussed, Vtail has no specialized domain fold but a set of interaction sites for acidic phospholipids, some of which are available in the autoinhibited conformation. Consistently, PIP_2 was suggested to promote vinculin targeting and activation in FA [106], but lipid binding-defective mutants of vinculin did not support this view. Instead, the vinculin/PIP_2 interaction was required for the inactivation of vinculin and adhesion site turnover. Two lipid binding-defective mutants of vinculin were studied in detail for their impact on FA dynamics. In one mutant, vinculin-ΔC, the C-terminal arm (residues 1053–1066) was deleted, and in the other, termed vinculin-LD (LD, lipid binding-deficient), positively charged amino acids, Arg and Lys, of the basic ladder and the basic collar in Vtail were exchanged for Gln [107, 108]. Compared to wild-type, Vtail constructs of both mutants displayed strongly reduced binding to acidic phospholipids, while F-actin binding and vinculin HTI were shown to be intact in vitro. For vinculin-LD, FRAP analysis gives identical half-life times of wild-type vinculin and the mutant, whereas residency times of Vhead constructs in FA are twice as long, again confirming that vinculin HTI is not affected [20, 108]. Upon expression in different cell lines, the lipid binding-defective mutants target efficiently and reveal no defect in protein localization, but FA turnover is strongly reduced. Vinculin-ΔC expression in vinculin null mouse embryonic fibroblasts [109] induces very stable adhesion sites with reduced turnover as compared to wild-type (re)expressing and vinculin-deficient control cells [107]. In B16-F1 melanoma cells, expression of vinculin-LD dominantly inhibits cell spreading as well as migration on extracellular matrix (2D) and in collagen gels (3D), in the presence of ~ 70% wild-type protein [108]. Together these results establish that Vtail binding to acidic phospholipids, or PIP_2 which is regulated in FA and therefore assumed to be the active compound, plays an important role in FA turnover.

Consistently, deregulated, high PIP_2 levels at the plasma membrane, for example induced by overexpression of phosphatidylinositol-4-phosphate 5 kinases, can lead to adhesion site dissolution and cell detachment. Expression of mouse PIPK1α, but not of a kinase dead mutant, induced rounding and detachment of B16-F1 cells, unless vinculin-LD was coexpressed and stabilized adhesion sites. The authors suggest that Vtail binding to the actin cytoskeleton is directly regulated by local PIP_2 levels. Thus, FA turnover in motile cells may dependent on a PIP_2-induced uncoupling of the cytoskeletal connection, with vinculin acting as sensor for PIP_2 levels [108].

Organism: Specific Tasks in Tissues and Organs

Talin and vinculin are highly conserved proteins in metazoans. In mouse, early lethality of knockout embryos reveals their unique function [109, 110], which is assumed also for many tissues and organs. Reconstitution experiments in knockout cells (derived from null embryos) provide an important tool to characterize key elements of protein function, but tissues/organs acquire further levels of regulation and complexity. Thus, verification of the main protein characteristics in genetic models is essential, and so far, a few selective mouse models have been established that address central aspects of talin and vinculin function.

For talin, its critical involvement in the activation of integrins was demonstrated in a mouse model for talin null platelets. These mice display severe hemostatic defects (bleeding) as a consequence of defective platelet activation. Upon stimulation, talin null platelets do not activate integrins, and platelet aggregation and adhesion to sites of vascular injury are not inducible either in vitro or in vivo, while other cellular functions of platelets are preserved [111, 112]. These results confirm that talin is an essential element in the regulation of integrins, but talin is not sufficient in platelets. As shown in further mouse models, talin acts in concert with the kindlin family of FERM domain proteins [113, 114].

For vinculin, the prominent localization in specialized adhesion structures of muscle cells, e.g. dense plaques of smooth muscles, and costameres and myotendinous junctions of striated muscle, suggested a critical involvement in cell junctions with specific adhesion properties and/or high mechanical load. Vinculin knockout embryos (vinc –/–) display defects in heart and brain development which, apart from the neural tube defect, were related rather to insufficient cell motility [109].

Instead, analysis of heterozygous adult animals (vinc wt/–), which are characterized by a roughly 50% reduction of vinculin protein in all tissues [S. Marg, W.H. Ziegler, unpubl. data], revealed that the organization of cell junctions in heart muscle, so-called intercalated disks, is disturbed, and upon cardiac stress, mice develop cardiomyopathies [115]. Consistently, the inducible, cardiomyocyte-specific knockout of vinculin in adult mice leads to disruption of cellular junctions in the heart, causing sudden death or dilated cardiomyopathy [116]. This mouse model confirms a critical contribution of vinculin to the structural and mechanical integrity of heart tissue, and supports other studies that have suggested a correlation of vinculin expression defects and mutations with cardiomyopathies [117–119].

At the organizational level of tissues and organs, further requirements for specialized structures and regulation apply, which may be satisfied by splice variants or isoforms of adaptor proteins, and/or specific interaction partners. The domain structure of the talin2 protein is very similar to that of talin1, however the *TLN2* gene structure is different and mRNA analysis revealed generation of multiple splice variants, which modulate aspects of talin rod function and may provide adapted, tissue-specific functionality, e.g. in brain and testis [86]. Likewise, muscle-specific expression of the vinculin splice variant metavinculin, with very similar biochemical properties and identical binding partners as vinculin, adds another level of complexity to adhesion site dynamics, which is not resolved mechanistically [120]. These protein variants, together with the still increasing number of interaction partners that often are not very well characterized and have not been in the focus of this article, are expected to provide the toolbox for a cell type and tissue-selective modulation of talin and vinculin ligand-binding activities.

Concluding Remarks

In the cell adhesion complex, investigation of protein function requires understanding of the structure, the biochemical and biophysical properties as well as the dynamic behavior of proteins, covering localization, state of activity, timing of interactions, protein levels and input signals. Proteins operate as entities, but modules of 'independent' functionality exist. Our two 'model' adaptor proteins, talin and vinculin, perform autoinhibitory interactions that are assumed to globally control binding activities. However, upon activation, domain folds may be released that can acquire independent functionality and modulate adhesion site regulation, like for example the talin head.

All proteins in the cell adhesion complex contribute to a transient, delicate equilibrium, balancing conservation of structures and force transmission on the one hand and adhesion site turnover and reorganization/motility on the other hand. Any deregulated, constitutive interaction can tip the balance, and by uncoupling turnover, force transmission and/or signaling, lead to inadequate behavior and inefficient motility of cells. Therefore, high-affinity interactions, as exemplified for talin and vinculin, are potentially detrimental and need cellular control at all relevant levels. The mutual dependence of ligand-binding activities makes analysis of protein domain constructs particularly demanding and requires adequate experimental control of adhesions site function in cells and in vivo.

Based on a fundamental understanding of the protein structure and the regulation of binding sites, genetic models in well-understood, 'simple' organisms like *Drosophila* and *Caenorhabditis elegans* can address the functional relevance of specific interactions and protein domains in tissues and organs. Tanentzapf and Brown [35, 121] have pioneered this approach for talin in *Drosophila*, showing that integrin activation by talin can be separated from localization and the mechanical connection, and furthermore, requires additional factors.

The discussion of selected adaptor proteins at the different organizational levels from molecular interactions to properties in tissues and organs leads to the conclusion that there is no simple, univocal answer to the question of adaptor protein function(s). Functional modules do not coincide with binding sites, and function being a consequence of the context, cannot be reduced to activities of individual domains that arrange like 'pearls on a string' in the protein. In fact, detailed information and further progress are needed at all levels and will contribute to a better understanding of core properties of tissue organization and cell motility in multicellular organisms, with relevance for development and regeneration, tissue homeostasis and states of disease.

Acknowledgements

I am grateful to the present and past members of my laboratory for their dedicated work and stimulating discussions. In addition, I am particularly indebted to Brigitte M. Jockusch, David R. Critchley and Reinhard Fässler for fostering my interest in the cytoskeletal field and long-term critical company and support. The work in the author's laboratory is supported by the Deutsche Forschungsgemeinschaft (DFG) and the Interdisciplinary Center for Clinical Research (IZKF) Leipzig.

References

1 Zaidel-Bar R, Itzkovitz S, Ma'ayan A, Iyengar R, Geiger B: Functional atlas of the integrin adhesome. Nat Cell Biol 2007;9:858–867.

2 Zamir E, Geiger B: Components of cell-matrix adhesions. J Cell Sci 2001;114:3577–3579.

3 Critchley DR: Focal adhesions – the cytoskeletal connection. Curr Opin Cell Biol 2000;12:133–139.

4 Ziegler WH, Liddington RC, Critchley DR: The structure and regulation of vinculin. Trends Cell Biol 2006;16:453–460.

5 Critchley DR: Biochemical and structural properties of the integrin-associated cytoskeletal protein talin. Annu Rev Biophys 2009;38:235–254.

6 Bakolitsa C, Cohen DM, Bankston LA, Bobkov AA, Cadwell GW, Jennings L, Critchley DR, Craig SW, Liddington RC: Structural basis for vinculin activation at sites of cell adhesion. Nature 2004;430:583–586.

7 Cohen DM, Chen H, Johnson RP, Choudhury B, Craig SW: Two distinct head-tail interfaces cooperate to suppress activation of vinculin by talin. J Biol Chem 2005;280:17109–17117.
8 Izard T, Evans G, Borgon RA, Rush CL, Bricogne G, Bois PR: Vinculin activation by talin through helical bundle conversion. Nature 2004;427:171–175.
9 Tran Van Nhieu G, Izard T: Vinculin binding in its closed conformation by a helix addition mechanism. EMBO J 2007;26:4588–4596.
10 Janssen ME, Kim E, Liu H, Fujimoto LM, Bobkov A, Volkmann N, Hanein D: Three-dimensional structure of vinculin bound to actin filaments. Mol Cell 2006;21:271–281.
11 Johnson RP, Craig SW: Actin activates a cryptic dimerization potential of the vinculin tail domain. J Biol Chem 2000;275:95–105.
12 Ramarao N, Le Clainche C, Izard T, Bourdet-Sicard R, Ageron E, Sansonetti PJ, Carlier MF, Tran Van Nhieu G: Capping of actin filaments by vinculin activated by the *Shigella* IpaA carboxyl-terminal domain. FEBS Lett 2007;581:853–857.
13 Steimle PA, Hoffert JD, Adey NB, Craig SW: Polyphosphoinositides inhibit the interaction of vinculin with actin filaments. J Biol Chem 1999;274: 18414–18420.
14 Niggli V: Structural properties of lipid-binding sites in cytoskeletal proteins. Trends Biochem Sci 2001; 26:604–611.
15 Johnson RP, Niggli V, Durrer P, Craig SW: A conserved motif in the tail domain of vinculin mediates association with and insertion into acidic phospholipid bilayers. Biochemistry 1998;37:10211–10222.
16 Bakolitsa C, de Pereda JM, Bagshaw CR, Critchley DR, Liddington RC: Crystal structure of the vinculin tail suggests a pathway for activation. Cell 1999; 99:603–613.
17 Ziegler WH, Tigges U, Zieseniss A, Jockusch BM: A lipid-regulated docking site on vinculin for protein kinase C. J Biol Chem 2002;277:7396–7404.
18 Palmer SM, Playford MP, Craig SW, Schaller MD, Campbell SL: Lipid binding to the tail domain of vinculin: specificity and the role of the N and C termini. J Biol Chem 2009;284:7223–7231.
19 Wood CK, Turner CE, Jackson P, Critchley DR: Characterisation of the paxillin-binding site and the C-terminal focal adhesion targeting sequence in vinculin. J Cell Sci 1994;107:709–717.
20 Humphries JD, Wang P, Streuli C, Geiger B, Humphries MJ, Ballestrem C: Vinculin controls focal adhesion formation by direct interactions with talin and actin. J Cell Biol 2007;179:1043–1057.
21 Goksoy E, Ma YQ, Wang X, Kong X, Perera D, Plow EF, Qin J: Structural basis for the autoinhibition of talin in regulating integrin activation. Mol Cell 2008; 31:124–133.
22 Goult BT, Bate N, Anthis NJ, Wegener KL, Gingras AR, Patel B, Barsukov IL, Campbell ID, Roberts GC, Critchley DR: The structure of an interdomain complex which regulates talin activity. J Biol Chem 2009;284:15097–15106.
23 Garcia-Alvarez B, de Pereda JM, Calderwood DA, Ulmer TS, Critchley D, Campbell ID, Ginsberg MH, Liddington RC: Structural determinants of integrin recognition by talin. Mol Cell 2003;11:49–58.
24 Wegener KL, Partridge AW, Han J, Pickford AR, Liddington RC, Ginsberg MH, Campbell ID: Structural basis of integrin activation by talin. Cell 2007;128:171–182.
25 Tadokoro S, Shattil SJ, Eto K, Tai V, Liddington RC, de Pereda JM, Ginsberg MH, Calderwood DA: Talin binding to integrin β tails: a final common step in integrin activation. Science 2003;302:103–106.
26 Cluzel C, Saltel F, Lussi J, Paulhe F, Imhof BA, Wehrle-Haller B: The mechanisms and dynamics of $\alpha_v\beta_3$ integrin clustering in living cells. J Cell Biol 2005;171:383–392.
27 Barsukov IL, Prescot A, Bate N, Patel B, Floyd DN, Bhanji N, Bagshaw CR, Letinic K, Di Paolo G, De Camilli P, Roberts GCK, Critchley DR: Phosphatidylinositol Phosphate kinase type 1γ and β_1-integrin cytoplasmic domain bind to the same region in the talin FERM domain. J Biol Chem 2003;278:31202–31209.
28 Wegener KL, Basran J, Bagshaw CR, Campbell ID, Roberts GC, Critchley DR, Barsukov IL: Structural basis for the interaction between the cytoplasmic domain of the hyaluronate receptor layilin and the talin F3 subdomain. J Mol Biol 2008;382:112–126.
29 Gingras AR, Ziegler WH, Bobkov AA, Joyce MG, Fasci D, Himmel M, Rothemund S, Ritter A, Grossmann JG, Patel B, Bate N, Goult BT, Emsley J, Barsukov IL, Roberts GC, Liddington RC, Ginsberg MH, Critchley DR: Structural determinants of integrin binding to the talin rod. J Biol Chem 2009; 284:8866–8876.
30 Tremuth L, Kreis S, Melchior C, Hoebeke J, Ronde P, Plancon S, Takeda K, Kieffer N: A fluorescence cell biology approach to map the second integrin-binding site of talin to a 130-amino-acid sequence within the rod domain. J Biol Chem 2004;279:22258–22266.
31 Rodius S, Chaloin O, Moes M, Schaffner-Reckinger E, Landrieu I, Lippens G, Lin M, Zhang J, Kieffer N: The talin rod IBS2 α-helix interacts with the β_3 integrin cytoplasmic tail membrane-proximal helix by establishing charge complementary salt bridges. J Biol Chem 2008;283:24212–24223.
32 Schmidt JM, Zhang J, Lee HS, Stromer MH, Robson RM: Interaction of talin with actin: sensitive modulation of filament crosslinking activity. Arch Biochem Biophys 1999;366:139–150.

33 Lee HS, Bellin RM, Walker DL, Patel B, Powers P, Liu H, Garcia-Alvarez B, de Pereda JM, Liddington RC, Volkmann N, Hanein D, Critchley DR, Robson RM: Characterization of an actin-binding site within the talin FERM domain. J Mol Biol 2004;343:771–784.

34 Hemmings L, Rees DJ, Ohanian V, Bolton SJ, Gilmore AP, Patel B, Priddle H, Trevithick JE, Hynes RO, Critchley DR: Talin contains three actin-binding sites each of which is adjacent to a vinculin-binding site. J Cell Sci 1996;109:2715–2726.

35 Tanentzapf G, Brown NH: An interaction between integrin and the talin FERM domain mediates integrin activation but not linkage to the cytoskeleton. Nat Cell Biol 2006;8:601–606.

36 Zhang X, Jiang G, Cai Y, Monkley SJ, Critchley DR, Sheetz MP: Talin depletion reveals independence of initial cell spreading from integrin activation and traction. Nat Cell Biol 2008;10:1062–1068.

37 Brett TJ, Legendre-Guillemin V, McPherson PS, Fremont DH: Structural definition of the F-actin-binding THATCH domain from HIP1R. Nat Struct Mol Biol 2006;13:121–130.

38 McCann RO, Craig SW: The I/LWEQ module: a conserved sequence that signifies F-actin binding in functionally diverse proteins from yeast to mammals. Proc Natl Acad Sci USA 1997;94:5679–5684.

39 Gingras AR, Bate N, Goult BT, Hazelwood L, Canestrelli I, Grossmann JG, Liu H, Putz NS, Roberts GC, Volkmann N, Hanein D, Barsukov IL, Critchley DR: The structure of the C-terminal actin-binding domain of talin. EMBO J 2008;27:458–469.

40 Smith SJ, McCann RO: A C-terminal dimerization motif is required for focal adhesion targeting of Talin1 and the interaction of the Talin1 I/LWEQ module with F-actin. Biochemistry 2007;46:10886–10898.

41 Srivastava J, Barreiro G, Groscurth S, Gingras AR, Goult BT, Critchley DR, Kelly MJ, Jacobson MP, Barber DL: Structural model and functional significance of pH-dependent talin-actin binding for focal adhesion remodeling. Proc Natl Acad Sci USA 2008;105:14436–14441.

42 Bass MD, Smith BJ, Prigent SA, Critchley DR: Talin contains three similar vinculin-binding sites predicted to form an amphipathic helix. Biochem J 1999;341:257–263.

43 Papagrigoriou E, Gingras AR, Barsukov IL, Bate N, Fillingham IJ, Patel B, Frank R, Ziegler WH, Roberts GC, Critchley DR, Emsley J: Activation of a vinculin-binding site in the talin rod involves rearrangement of a 5-helix bundle. EMBO J 2004;23:2942–2951.

44 Gingras AR, Ziegler WH, Frank R, Barsukov IL, Roberts GC, Critchley DR, Emsley J: Mapping and consensus sequence identification for multiple vinculin binding sites within the talin rod. J Biol Chem 2005;280:37217–37224.

45 Gingras AR, Vogel KP, Steinhoff HJ, Ziegler WH, Patel B, Emsley J, Critchley DR, Roberts GC, Barsukov IL: Structural and dynamic characterization of a vinculin binding site in the talin rod. Biochemistry 2006;45:1805–1817.

46 Patel B, Gingras AR, Bobkov AA, Fujimoto LM, Zhang M, Liddington RC, Mazzeo D, Emsley J, Roberts GC, Barsukov IL, Critchley DR: The activity of the vinculin binding sites in talin is influenced by the stability of the helical bundles that make up the talin rod. J Biol Chem 2006;281:7458–7467.

47 Hytonen VP, Vogel V: How force might activate talin's vinculin binding sites: smd reveals a structural mechanism. PLoS Comput Biol 2008;4:e24.

48 Lee SE, Chunsrivirot S, Kamm RD, Mofrad MR: Molecular dynamics study of talin-vinculin binding. Biophys J 2008;95:2027–2036.

49 Del Rio A, Perez-Jimenez R, Liu R, Roca-Cusachs P, Fernandez JM, Sheetz MP: Stretching single talin rod molecules activates vinculin binding. Science 2009;323:638–641.

50 Galbraith CG, Yamada KM, Sheetz MP: The relationship between force and focal complex development. J Cell Biol 2002;159:695–705.

51 Gallant ND, Michael KE, Garcia AJ: Cell adhesion strengthening: contributions of adhesive area, integrin binding, and focal adhesion assembly. Mol Biol Cell 2005;16:4329–4340.

52 Fillingham I, Gingras AR, Papagrigoriou E, Patel B, Emsley J, Critchley DR, Roberts GC, Barsukov IL: A vinculin binding domain from the talin rod unfolds to form a complex with the vinculin head. Structure (Camb) 2005;13:65–74.

53 Wouters FS, Verveer PJ, Bastiaens PI: Imaging biochemistry inside cells. Trends Cell Biol 2001;11:203–211.

54 Bulinski JC, Odde DJ, Howell BJ, Salmon TD, Waterman-Storer CM: Rapid dynamics of the microtubule binding of ensconsin in vivo. J Cell Sci 2001;114:3885–3897.

55 Sprague BL, McNally JG: FRAP analysis of binding: proper and fitting. Trends Cell Biol 2005;15:84–91.

56 Bunt G, Wouters FS: Visualization of molecular activities inside living cells with fluorescent labels. Int Rev Cytol 2004;237:205–277.

57 Papusheva E, Mello de Queiroz F, Dalous J, Han Y, Esposito A, Jares-Erijmanxa EA, Jovin TM, Bunt G: Dynamic conformational changes in the FERM domain of FAK are involved in focal-adhesion behavior during cell spreading and motility. J Cell Sci 2009;122:656–666.

58 Himmel M, Ritter A, Rothemund S, Pauling BV, Rottner K, Gingras AR, Ziegler WH: Control of high-affinity interactions in the talin C terminus: how talin domains coordinate protein dynamics in cell adhesions. J Biol Chem 2009;284:13832–13842.
59 Chen H, Cohen DM, Choudhury DM, Kioka N, Craig SW: Spatial distribution and functional significance of activated vinculin in living cells. J Cell Biol 2005;169:459–470.
60 Cohen DM, Kutscher B, Chen H, Murphy DB, Craig SW: A conformational switch in vinculin drives formation and dynamics of a talin-vinculin complex at focal adhesions. J Biol Chem 2006;281:16006–16015.
61 Chen H, Choudhury DM, Craig SW: Coincidence of actin filaments and talin is required to activate vinculin. J Biol Chem 2006;281:40389–40398.
62 Moes M, Rodius S, Coleman SJ, Monkley SJ, Goormaghtigh E, Tremuth L, Kox C, van der Holst PP, Critchley DR, Kieffer N: The integrin binding site 2 in the talin rod domain is essential for linking integrin β subunits to the cytoskeleton. J Biol Chem 2007;282:17280–17288.
63 Subauste MC, Pertz O, Adamson ED, Turner CE, Junger S, Hahn KM: Vinculin modulation of paxillin-FAK interactions regulates ERK to control survival and motility. J Cell Biol 2004;165:371–381.
64 Mierke CT, Kollmannsberger P, Zitterbart DP, Smith J, Fabry B, Goldmann WH: Mechano-coupling and regulation of contractility by the vinculin tail domain. Biophys J 2008;94:661–670.
65 Coll JL, Ben-Ze'ev A, Ezzell RM, Rodriguez Fernandez JL, Baribault H, Oshima RG, Adamson ED: Targeted disruption of vinculin genes in F9 and embryonic stem cells changes cell morphology, adhesion, and locomotion. Proc Natl Acad Sci USA 1995;92:9161–9165.
66 Volberg T, Geiger B, Kam Z, Pankov R, Simcha I, Sabanay H, Coll JL, Adamson E, Ben-Ze'ev A: Focal adhesion formation by F9 embryonal carcinoma cells after vinculin gene disruption. J Cell Sci 1995; 108:2253–2260.
67 Xu W, Coll JL, Adamson ED: Rescue of the mutant phenotype by reexpression of full-length vinculin in null F9 cells; effects on cell locomotion by domain deleted vinculin. J Cell Sci 1998;111:1535–1544.
68 Hong T, Grabel LB: Migration of F9 parietal endoderm cells is regulated by the ERK pathway. J Cell Biochem 2006;97:1339–1349.
69 Klemke RL, Cai S, Giannini AL, Gallagher PJ, de Lanerolle P, Cheresh DA: Regulation of cell motility by mitogen-activated protein kinase. J Cell Biol 1997; 137:481–492.
70 Glading A, Bodnar RJ, Reynolds IJ, Shiraha H, Satish L, Potter DA, Blair HC, Wells A: Epidermal growth factor activates m-calpain (calpain II), at least in part, by extracellular signal-regulated kinase-mediated phosphorylation. Mol Cell Biol 2004;24:2499–2512.
71 Webb DJ, Donais K, Whitmore LA, Thomas SM, Turner CE, Parsons JT, Horwitz AF: FAK-Src signalling through paxillin, ERK and MLCK regulates adhesion disassembly. Nat Cell Biol 2004;6:154–161.
72 Franco SJ, Huttenlocher A: Regulating cell migration: calpains make the cut. J Cell Sci 2005;118:3829–3838.
73 Mierke CT, Rosel D, Fabry B, Brabek J: Contractile forces in tumor cell migration. Eur J Cell Biol 2008; 87:669–676.
74 Mierke CT: The role of vinculin in the regulation of the mechanical properties of cells. Cell Biochem Biophys 2009;53:115–126.
75 Zamir E, Geiger B: Molecular complexity and dynamics of cell-matrix adhesions. J Cell Sci 2001; 114:3583–3590.
76 Zaidel-Bar R, Ballestrem C, Kam Z, Geiger B: Early molecular events in the assembly of matrix adhesions at the leading edge of migrating cells. J Cell Sci 2003;116:4605–4613.
77 Nobes CD, Hall A: Rho, rac, and cdc42 GTPases regulate the assembly of multimolecular focal complexes associated with actin stress fibers, lamellipodia, and filopodia. Cell 1995;81:53–62.
78 Rottner K, Hall A, Small JV: Interplay between Rac and Rho in the control of substrate contact dynamics. Curr Biol 1999;9:640–648.
79 Giannone G, Jiang G, Sutton DH, Critchley DR, Sheetz MP: Talin1 is critical for force-dependent reinforcement of initial integrin-cytoskeleton bonds but not tyrosine kinase activation. J Cell Biol 2003; 163:409–419.
80 Schlaepfer DD, Hauck CR, Sieg DJ: Signaling through focal adhesion kinase. Prog Biophys Mol Biol 1999;71:435–478.
81 Lele TP, Pendse J, Kumar S, Salanga M, Karavitis J, Ingber DE: Mechanical forces alter zyxin unbinding kinetics within focal adhesions of living cells. J Cell Physiol 2006;207:187–194.
82 Brown CM, Hebert B, Kolin DL, Zareno J, Whitmore L, Horwitz AR, Wiseman PW: Probing the integrin-actin linkage using high-resolution protein velocity mapping. J Cell Sci 2006;119:5204–5214.
83 Hu K, Ji L, Applegate KT, Danuser G, Waterman-Storer CM: Differential transmission of actin motion within focal adhesions. Science 2007;315: 111–115.

84 Von Wichert G, Haimovich B, Feng GS, Sheetz MP: Force-dependent integrin-cytoskeleton linkage formation requires downregulation of focal complex dynamics by Shp2. EMBO J 2003;22:5023–5035.
85 Ratnikov B, Ptak C, Han J, Shabanowitz J, Hunt DF, Ginsberg MH: Talin phosphorylation sites mapped by mass spectrometry. J Cell Sci 2005;118:4921–4923.
86 Debrand E, El Jai Y, Spence L, Bate N, Praekelt U, Pritchard CA, Monkley SJ, Critchley DR: Talin2 is a large and complex gene encoding multiple transcripts and protein isoforms. FEBS J 2009;276:1610–1628.
87 Borowsky ML, Hynes RO: Layilin, a novel talin-binding transmembrane protein homologous with C-type lectins, is localized in membrane ruffles. J Cell Biol 1998;143:429–442.
88 Michael KE, Dumbauld DW, Burns KL, Hanks SK, Garcia AJ: Focal adhesion kinase modulates cell adhesion strengthening via integrin activation. Mol Biol Cell 2009;20:2508–2519.
89 Zhang Z, Izaguirre G, Lin SY, Lee HY, Schaefer E, Haimovich B: The phosphorylation of vinculin on tyrosine residues 100 and 1065, mediated by SRC kinases, affects cell spreading. Mol Biol Cell 2004; 15:4234–4247.
90 Mohl C, Kirchgessner N, Schafer C, Kupper K, Born S, Diez G, Goldmann WH, Merkel R, Hoffmann B: Becoming stable and strong: the interplay between vinculin exchange dynamics and adhesion strength during adhesion site maturation. Cell Motil Cytoskeleton 2009;66:350–364.
91 Zaidel-Bar R, Milo R, Kam Z, Geiger B: A paxillin tyrosine phosphorylation switch regulates the assembly and form of cell-matrix adhesions. J Cell Sci 2007;120:137–148.
92 Wolfenson H, Lubelski A, Regev T, Klafter J, Henis YI, Geiger B: A role for the juxtamembrane cytoplasm in the molecular dynamics of focal adhesions. PLoS ONE 2009;4:e4304.
93 Martel V, Racaud-Sultan C, Dupe S, Marie C, Paulhe F, Galmiche A, Block MR, Albiges-Rizo C: Conformation, localization and integrin-binding of talin depend on its interaction with phosphoinositides. J Biol Chem 2001;276:21217–21227.
94 Bos JL: Linking Rap to cell adhesion. Curr Opin Cell Biol 2005;17:123–128.
95 Han J, Lim CJ, Watanabe N, Soriani A, Ratnikov B, Calderwood DA, Puzon-McLaughlin W, Lafuente EM, Boussiotis VA, Shattil SJ, Ginsberg MH: Reconstructing and deconstructing agonist-induced activation of integrin αIIbβ3. Curr Biol 2006;16: 1796–1806.
96 Lee HS, Lim CJ, Puzon-McLaughlin W, Shattil SJ, Ginsberg MH: RIAM activates integrins by linking talin to ras GTPase membrane-targeting sequences. J Biol Chem 2009;284:5119–5127.
97 Watanabe N, Bodin L, Pandey M, Krause M, Coughlin S, Boussiotis VA, Ginsberg MH, Shattil SJ: Mechanisms and consequences of agonist-induced talin recruitment to platelet integrin αIIbβ3. J Cell Biol 2008;181:1211–1222.
98 Di Paolo G, Pellegrini L, Letinic K, Cestra G, Zoncu R, Voronov S, Chang S, Guo J, Wenk MR, De Camilli P: Recruitment and regulation of phosphatidylinositol phosphate kinase type 1γ by the FERM domain of talin. Nature 2002;420:85–89.
99 Ling K, Doughman RL, Firestone AJ, Bunce MW, Anderson RA: Type I γ-phosphatidylinositol phosphate kinase targets and regulates focal adhesions. Nature 2002;420 89–93.
100 Ling K, Doughman RL, Iyer VV, Firestone AJ, Bairstow SF, Mosher DF, Schaller MD, Anderson RA: Tyrosine phosphorylation of type I γ-phosphatidylinositol phosphate kinase by Src regulates an integrin-talin switch. J Cell Biol 2003;163: 1339–1349.
101 Franco SJ, Rodgers MA, Perrin BJ, Han J, Bennin DA, Critchley DR, Huttenlocher A: Calpain-mediated proteolysis of talin regulates adhesion dynamics. Nat Cell Biol 2004;6:977–983.
102 Huang C, Rajfur Z, Yousefi N, Chen Z, Jacobson K, Ginsberg MH: Talin phosphorylation by Cdk5 regulates Smurf1-mediated talin head ubiquitylation and cell migration. Nat Cell Biol 2009;11:624–630.
103 Bhatt A, Kaverina I, Otey C, Huttenlocher A Regulation of focal complex composition and disassembly by the calcium-dependent protease calpain. J Cell Sci 2002;115:3415–3425.
104 Serrano K, Devine DV: Vinculin is proteolyzed by calpain during platelet aggregation: 95 kDa cleavage fragment associates with the platelet cytoskeleton. Cell Motil Cytoskeleton 2004;58:242–252.
105 Weber H, Huhns S, Luthen F, Jonas L: Calpain-mediated breakdown of cytoskeletal proteins contributes to cholecystokinin-induced damage of rat pancreatic acini. Int J Exp Pathol 2009;90:387–399.
106 Gilmore AP, Burridge K: Regulation of vinculin binding to talin and actin by phosphatidyl-inositol-4,5-bisphosphate. Nature 1996;381:531–535.
107 Saunders RM, Holt MR, Jennings L, Sutton DH, Barsukov IL, Bobkov A, Liddington RC, Adamson EA, Dunn GA, Critchley DR: Role of vinculin in regulating focal adhesion turnover. Eur J Cell Biol 2006;85:487–500.
108 Chandrasekar I, Stradal TE, Holt MR, Entschladen F, Jockusch BM, Ziegler WH: Vinculin acts as a sensor in lipid regulation of adhesion-site turnover. J Cell Sci 2005;118:1461–1472.

109 Xu W, Baribault H, Adamson ED: Vinculin knockout results in heart and brain defects during embryonic development. Development 1998;125:327–337.
110 Monkley SJ, Zhou XH, Kinston SJ, Giblett SM, Hemmings L, Priddle H, Brown JE, Pritchard CA, Critchley DR, Fassler R: Disruption of the talin gene arrests mouse development at the gastrulation stage. Dev Dyn 2000;219:560–574.
111 Nieswandt B, Moser M, Pleines I, Varga-Szabo D, Monkley S, Critchley D, Fassler R: Loss of talin1 in platelets abrogates integrin activation, platelet aggregation, and thrombus formation in vitro and in vivo. J Exp Med 2007;204:3113–3118.
112 Petrich BG, Marchese P, Ruggeri ZM, Spiess S, Weichert RA, Ye F, Tiedt R, Skoda RC, Monkley SJ, Critchley DR, Ginsberg MH: Talin is required for integrin-mediated platelet function in hemostasis and thrombosis. J Exp Med 2007;204:3103–3111.
113 Moser M, Nieswandt B, Ussar S, Pozgajova M, Fassler R: Kindlin-3 is essential for integrin activation and platelet aggregation. Nat Med 2008;14:325–330.
114 Moser M, Legate KR, Zent R, Fassler R: The tail of integrins, talin, and kindlins. Science 2009;324:895–899.
115 Zemljic-Harpf AE, Ponrartana S, Avalos RT, Jordan MC, Roos KP, Dalton ND, Phan VQ, Adamson ED, Ross RS: Heterozygous inactivation of the vinculin gene predisposes to stress-induced cardiomyopathy. Am J Pathol 2004;165:1033–1044.
116 Zemljic-Harpf AE, Miller JC, Henderson SA, Wright AT, Manso AM, Elsherif L, Dalton ND, Thor AK, Perkins GA, McCulloch AD, Ross RS: Cardiac-myocyte-specific excision of the vinculin gene disrupts cellular junctions, causing sudden death or dilated cardiomyopathy. Mol Cell Biol 2007;27:7522–7537.
117 Maeda M, Holder E, Lowes B, Valent S, Bies RD: Dilated cardiomyopathy associated with deficiency of the cytoskeletal protein metavinculin. Circulation 1997;95:17–20.
118 Olson TM, Illenberger S, Kishimoto NY, Huttelmaier S, Keating MT, Jockusch BM: Metavinculin mutations alter actin interaction in dilated cardiomyopathy. Circulation 2002;105:431–437.
119 Vasile VC, Ommen SR, Edwards WD, Ackerman MJ: A missense mutation in a ubiquitously expressed protein, vinculin, confers susceptibility to hypertrophic cardiomyopathy. Biochem Biophys Res Commun 2006;345:998–1003.
120 Witt S, Zieseniss A, Fock U, Jockusch BM, Illenberger S: Comparative biochemical analysis suggests that vinculin and metavinculin cooperate in muscular adhesion sites. J Biol Chem 2004;279: 31533–31543.
121 Tanentzapf G, Martin-Bermudo MD, Hicks MS, Brown NH: Multiple factors contribute to integrin-talin interactions in vivo. J Cell Sci 2006;119:1632–1644.

Dr. Wolfgang H. Ziegler
Interdisciplinary Center for Clinical Research (IZKF) Leipzig
Faculty of Medicine, University of Leipzig
Inselstrasse 22, DE–04103 Leipzig (Germany)
Tel. +49 341 971 5945, Fax +49 341 971 5979, E-Mail wolfgang.ziegler@medizin.uni-leipzig.de

Entschladen F, Zänker KS (eds): Cell Migration: Signalling and Mechanisms.
Transl Res Biomed. Basel, Karger, 2010, vol 2, pp 163–172

Locomotor Force Generation by Myosins

Jbireal M. Ali Jbireal · Frank Entschladen · Kurt S. Zänker

Institute of Immunology, Witten/Herdecke University, Witten, Germany

Abstract

Similar to muscle contraction, the generation of locomotor forces in migrating cells largely depends on the interaction of myosin and actin. Besides non-muscle myosin II being the most important isotype for migration, several other myosins have been described to be involved in cell migration. The activity of these myosins in migrating cells is regulated by the phosphorylation of the regulatory light chain. We herein provide insight into the intracellular signal transduction pathways that regulate this phosphorylation of myosin and summarize the current knowledge on the involvement of the various myosin isotypes in migratory processes.

The Myosin Superfamily

As mentioned in the previous chapter, the interaction of actin and myosin is an essential event not only for muscle contraction but also for the movement of single eukaryotic cells. The role of actin in cell migration has been elucidated in detail in the past decades, as discussed above, but considerably less attention has been paid to the involvement and regulation of myosin. However, although there is still a lack of molecular data, some cell migration models consider myosin to play a crucial role in the generation of migration forces [1–3].

Myosins are defined as actin-dependent molecular motors that use the energy of ATP hydrolysis to move along actin filaments. Myosins consist of a heavy chain and attached light chains. The heavy chain has a conserved tripartite domain organization consisting of a globular catalytic domain of about 80 kDa size, also called the motor domain that binds actin and has ATPase activity. The motor domain is followed by a neck region containing one to six IQ motifs (with the consensus sequence IQXXXRGXXXR) that binds myosin light chains or calmodulin, and a highly variable carboxy-terminal tail domain essential for cargo binding [4]. The carboxy-terminal tail domain is the most diverse part of myosin, ranging from very long to rather short [5]. Some myosins are dimeric molecules, including the isotypes of conventional motor protein myosin II, which is the only family that forms bipolar bundles by self-assembly of the tail region [5, 6]. The classification into the 18 families is based on differences in the amino-acid sequence of the motor domain, however, myosins within each class are also similar in terms of their tail domain organization [4]. Presently, 18 myosin families with about 140 members are known, and 40 myosin genes belonging to 12 of these families are expressed in humans [6, 7], whereas the expression of some myosins is tissue-specific [7].

There are two types of myosin light chains, the essential light chain, which has solely structural function, and the regulatory light chain (RLC), which has regulatory function for the motor domain [8]. Phosphorylation of the RLC leads to conformational changes of the whole myosin complex, which then allows binding of the motor domain to actin [8]. This phosphorylation is a key regulatory event in smooth muscle and non-muscle cells, whereas only in skeletal muscle cells the troponin-tropomyosin complex regulates the myosin-actin interaction [9]. Although the RLC is phosphorylated in skeletal muscle cells, the non-phosphorylated RLC is not able to inhibit the activity of vertebrate skeletal muscle myosin [10]. The phosphorylation of the RLC, also termed as myosin light chain (MLC) is regulated by two kinases, the calcium-dependent myosin light chain kinase (MLCK) and some calcium-independent kinases, i.e. the Rho-kinase/Rho-dependent coiled-coil kinase (ROCK), the integrin-linked kinase (ILK), the p21-activated protein kinase (PAK), and others [11, 12] (fig. 1). The MLC is dephosphorylated by the myosin light chain phosphatase (MLCP), for which some of aforementioned kinases are negative regulators, too [11, 12]. However, calcium regulation is supposed to be the most important signal for the contraction of vascular smooth muscle, as it is not only a regulator for the MLCK via calmodulin, but also for further signaling pathways which regulate MLCP activity [11].

It has already been demonstrated in the previous chapters that G-protein-coupled receptors (GPCRs, also known as serpentine receptors or seven-helices (7H) receptors) probably constitute the most important, but not the only family of receptors for signal substances which regulate the migratory activity of cells and which are furthermore able to induce chemotaxis. The most prominent ligands to GPCRs with regard to migration are chemokines and neurotransmitters [13, 14]. GPCRs activate upon ligand binding on the intracellular side the name-giving heterotrimeric G proteins. These proteins spilt into a GTP-binding α subunit, and a βγ subunit. The schematic drawing (fig. 1) displays pathways that are activated by G_s-coupled receptors, which means that the α subunit stimulates the adenylyl cyclase. There are further classes of α subunits which have distinct targets, i.e. $G_{i/o}$, $G_{q/11}$, and $G_{12/13}$ proteins [15]. However, the type of G protein which is activated by each receptor and its according ligand are individual and very often there is more than one receptor for each ligand, which are coupled to different G proteins. For example, there are several serotonin receptors (5-HT_1 to $5HT_7$) which are – with the exception of 5-HT_3 – GPCRs that are coupled to G_s, $G_{i/o}$, and $G_{q/11}$ proteins [16]. With regard to the regulation of myosin activity, two signaling events seem to be of special importance. Firstly, the βγ subunit of the G proteins activates either the phospholipase Cγ (PLCγ) via G protein-coupled receptor kinases (GRK) [17], and subsequently via β-arrestin and protein tyrosine kinases

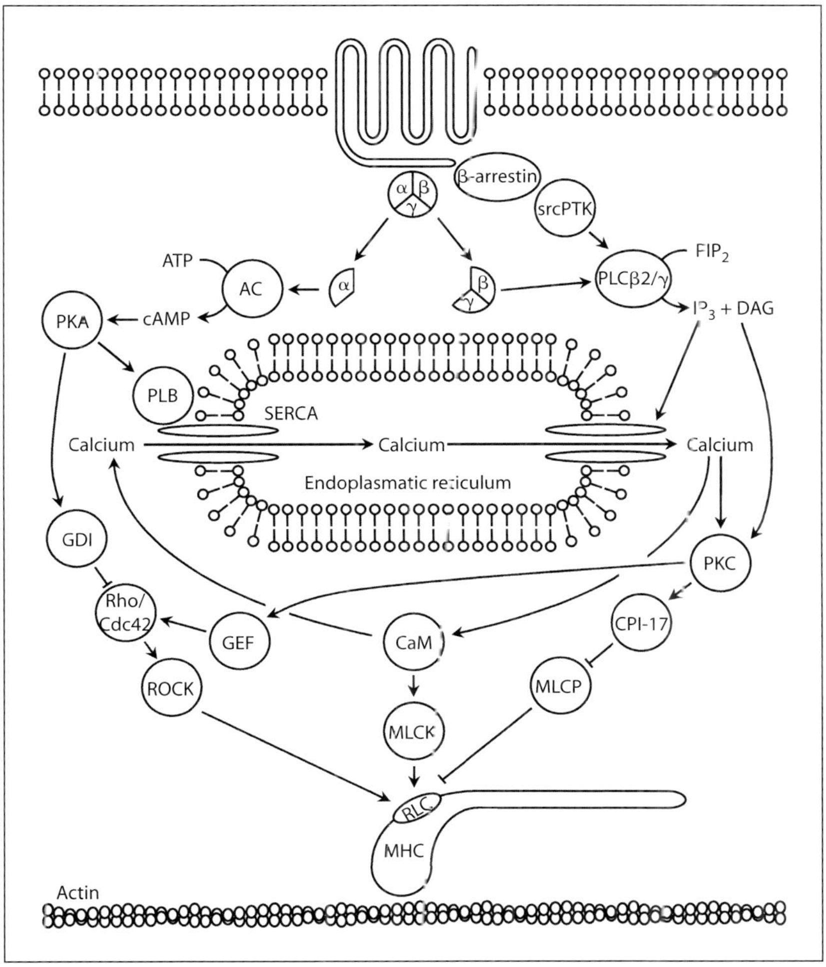

Fig. 1. Schematic drawing of the regulation of myosin activity in migrating cells.

(PTK) [18], or activates the PLCβ_2 by direct interaction [19]. The PLCs generate two second messengers, diacylglycerol (DAG) and inositol-trisphosphate (IP_3), by the cleavage of phosphatidylinositol-bisphosphate (PIP_2) [20]. DAG, in concert with calcium, is an activator for so-called classical isotypes of the protein kinase C (PKC) [21]. IP_3 opens intracellular calcium channels and thus leads to a rapid increase of cytosolic calcium levels [22]. This is a key event for the activation of myosin. Calcium binds to calmodulin, which activates the MLCK [12]. In addition, the PKC activates Rho [23], probably via the activation of GDP/GTP exchange factors (GEF), whereas it is also under discussion that the GEFs are directly activated by DAG [24]. Rho in turn activates ROCK [11, 12]. The second signaling pathway is complementary and in parts antagonistic to the first one: the α subunit actives the adenylyl cyclase (AC) which then generates cyclic AMP (cAMP). This second messenger is an activator for the protein kinase A (PKA) [25]. The PKA phosphorylates phospholamban (PLB) which then dissociates from the sarcoplasmatic/endoplasmatic calcium ATPase (SERCA). This leads to a relatively slow calcium uptake into the endoplasmatic reticulum. Furthermore, the PKA inhibits Rho via activation of GDP dissociation inhibitors (GDI) [26]. Thus, Rho is under the dual control of the PKA and PKC [27]. In addition to these two pathways, which regulate the activators of myosin, GPCRs inhibit the MLCP by partially the same pathway (fig. 1). The PKC phosphorylates the inhibitory protein CPI-17 (17-kDa PKC-potentiated inhibitory protein of type 1 protein phosphatase), which then has inhibitory function for the MLCP [11].

Although all myosins are motor proteins with an overall common structure and regulation, their function in detail is very diverse. We herein present the current knowledge on the involvement of the various myosin isotypes in the locomotor force generation of different cell type and other cellular motion processes.

Myosin I

The first unconventional myosin discovered was the low molecular weight motor protein called myosin I, which was purified from *Acanthamoeba castellanii* [28]. Much of our understanding of the biochemical and functional properties of this class of myosin derives from studies of the amoeboid organisms *Acanthamoeba* and *Dictyostelium* [29]. All characterized type I myosins are low-duty ratio motors, and the binding states have a very low actin affinity. Therefore, a small amount of myosin I proteins cannot support contraction or movements of cargo over long distances [30]. Studies of three myosin I isotypes (myoIA, myoIB, and myoIF) in *Dictyostelium discoideum* implicate single-headed, non-filament-forming myosin I variants in the formation of pseudopods. This role is accomplished via a mechanism that is different from that employing myosin II. Myosin I mutants in *Dictyostelium* exhibit a decreased velocity of chemotaxis due to increased formation of lateral pseudopods [31]. Furthermore, excess full-length cortical myoIB expression in *Dictyostelium* prevents the formation of the actin-filled extensions required for the locomotion by increasing the tension of the filamentous (F)-actin cytoskeleton and/or retracting projections before they can fully extend [32]. The myoIB is a calmodulin-associated protein widely expressed in mammalian tissues, which is supposed to have actin cross-linking function through a cryptic actin-binding site [33].

Myosin II

Myosin II isoforms are the major contractile proteins in muscles and also play a crucial role in non-muscle cell contractility. Myosin II molecules contain two motor domains and assemble into bipolar filaments [33]. Like the myosin I family, phylogenetic analysis of the motor domains of the myosin II family shows a division of isoforms into distinct subclasses that cross phylogenetic boundaries [6]. As already described above, myosin II isoforms are the only myosins that from bipolar bundles by tail assembly [5, 6]. To date, three isoforms of non-muscle myosin heavy chains (NMHC) II, termed NMHC II-A, NMHC II-B and NMHC II-C, have been identified in vertebrates [34, 35]. In striated, skeletal muscle cells, the troponin-tropomyosin complex regulates the interaction of myosin II with actin [9]. In smooth muscle and non-muscle cells, this complex is not expressed and regulation occurs by the above described pathways (fig. 1). Recent studies, however, demonstrate that the activity and regulation of non-muscle myosin II may differ significantly from that of smooth muscle myosin II. For instance, cytokinesis and polarized cell locomotion require the correct temporal and spatial regulation of myosin II and associated cytoskeletal proteins, whereas in adult smooth muscle there is little remodeling of actomyosin assembles [36].

Non-muscle myosin II is a key motor protein in the migration of leukocytes and tumor cells. Recent investigations of non-muscle myosin II demonstrate that it is relatively high-duty ratio motor under physiological actin and nucleotide concentrations; myosin is attached to actin for about 20–50% of its ATPase cycle time [37]. In neutrophil granulocytes and cytotoxic T lymphocytes, migratory activity is almost abrogated by the inhibition of myosin II activity, whereas in carcinoma cells of breast and prostate origin, only the migratory activity that was induced by GPCRs was affected by inhibition of myosin II activity [38]. Myosin II is important for the uropod contraction in neutrophil granulocytes [39]. However, other models that consider differences between two- and three-dimensional migration assays regard myosin II important for the migration of leukocytes only in certain situations as through narrow gaps [40]. Macrophage migration was only myosin II-dependent in three-dimensional systems, whereas it was independent of myosin II in two-dimensional migration assays [41]. It is thus an important matter which kind of migration assay is used, not only for myosin involvement but for further aspects of molecular migration mechanisms as well [38, 42].

Myosin V

There are three myosin V isoforms expressed in humans [5], with a tissue-specific distribution of expression [43]. Type V myosins are double-headed and calcium/calmodulin-regulated molecules mainly involved in the transport of vesicles and organelles [7, 44]. Especially the multiple functions of myosin Va in various cells and tissues are well characterized [45]. Its functions range from particle uptake in macrophages [46] to the melanosome transport in melanocytes [47]. Furthermore myosin Va is supposed to play a role in cancer cell migration and metastasis formation, whereas the expression of myosin Va seems to be under the control of snail, a regulator for the epithelial-mesenchymal transition [48].

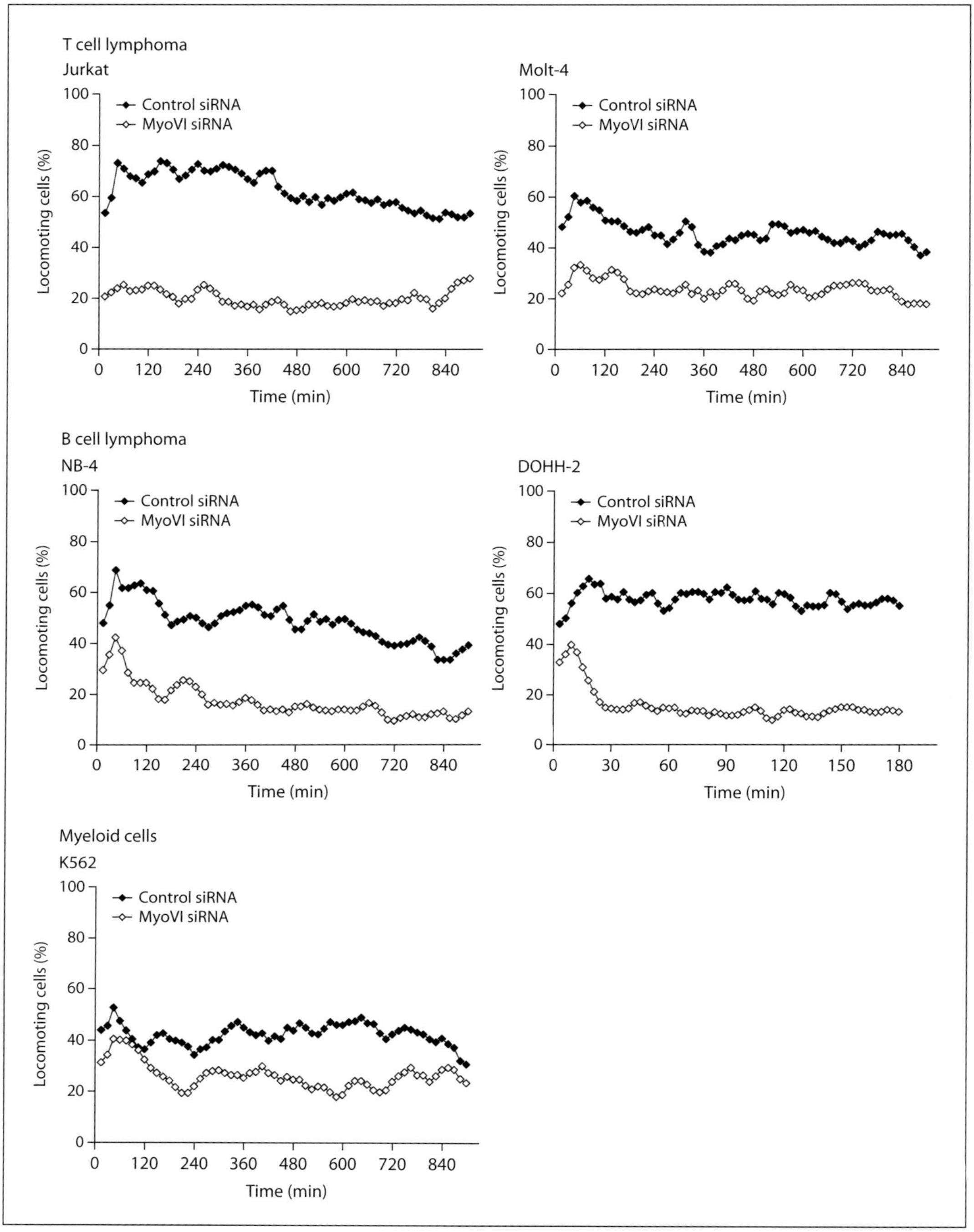

Fig. 2. Migratory activity of cells from leukemic cell lines. The cells were embedded with a three-dimensional collagen matrix [42]. The migratory activity was recorded by time-lapse videomicroscopy and analyzed by computer-assisted cell tracking. The graphs show mean values of three independent experiments. Please note that the time scale concerning the DOHH-2 cells is different from all other cells, since these cells migrate much faster. All cells were treated with unspecific control siRNA, or siRNA that specifically downregulates the myosin VI expression, as was controlled by immunoblotting.

Myosin VI

In mammalians, four isoforms from one myosin VI gene are expressed, which are splice variants with inserts in the tail region [49]. The isoforms have either one of two different inserts, both, or none, whereas the expression pattern of the isoforms is tissue-specific [49]. Myosin VI is of particular nature, since it is only one of two myosins known to move towards the pointed (minus) end of the actin filament, instead of the barbed (positive) end, as the other myosins [50].

Myosin VI is involved in endocytotic processes; it binds to partners of clathrin-coated pits as well as of uncoated endocytotic vesicles [49]. Furthermore, myosin VI plays a role in hearing. Mutations of myosin VI in mice (known as Schnell's waltzer mice) [5]. With regard to cell migration, myosin VI is supposed to play a role in several cell types. In *Drosophila melanogaster*, myosin VI is responsible for membrane protrusion at the leading edge the in the migration of border cells in the ovary [51]. In humans, myosin VI is involved in the migration of fibroblasts [52], ovarian cancer cells [53], and prostate cancer [54]. In the latter case, overexpression of myosin VI was correlated with aggressive histological features and the presence of the androgen receptor. Our own results show that myosin VI is critically involved in the migration of cells from myeloid and lymphoid leukemic cell lines (fig. 2).

Myosin VII

Myosin VII is an important mediator of cell adhesion. It is required for the initial stages of the particle adhesion during phagocytosis and cell surface adhesion during migration of *Dictyostelium* [55]. In these organisms, myosin VII interacts with the actin-binding protein talinA [56]. In humans, two isoforms of myosin VII are expressed, myosin VIIa and VIIb, which share nearly 50% homology [5]. Mutations of myosin VIIa are linked to the USHER1B syndrome, which impairs hearing and seeing [5]; over 40 mutations of the myosin VIIa gene are known [57]. However, to our knowledge, a role of myosin VII in human cell migration has not been proven so far. We have detected an expression of myosin VIIa in neutrophil granulocytes, but not in cytotoxic T lymphocytes as well as in cell lines of breast and prostate carcinoma.

Myosin X

One myosin X is expressed in humans [6] which is supposed to be involved in the pseudopod extension during phagocytosis and filopodia extension during migration [7, 58], where it is involved in β-integrin transportation [59]. Myosin X has three plekstrin homology domains [60] which link this myosin to GPCR signaling via phosphoinositides.

Table 1. Current knowledge on which myosin family members are involved in the migration of various cells

Isotype	Function in humans and other organisms
Myosin I	– Pseudopod formation in *Dictyostelium* – myoIb has an actin cross-linking function, widely expressed in mammalians, involvement in human cell migration not proven yet
Myosin II	– Crucial for leukocyte migration and GPCR-induced carcinoma cell migration
Myosin Va	– Macrophage phagocytosis – Melanosome transport in melanocytes – Epithelial-mesenchymal transition in cancer cells
Myosin VI	– Border cell migration in *Drosophila* – Endocytosis – Migration of fibroblasts, ovarian cancer, prostate cancer, leukemic cells
Myosin VIIa	– Surface adhesion in *Dictyostelium* migration – No function in human cell migration shown yet

Concluding Remarks

Not only myosin II, but also several other myosin isotypes seem to be involved in cell migration processes, whereas there is a certain myosin pattern for each cell type. Table 1 summarizes the current knowledge on which myosin family members are involved in the migration of various cells.

Acknowledgement

This work was supported by the Fritz Bender Foundation (Munich, Germany).

References

1 Charras G, Paluch E: Blebs lead the way: how to migrate without lamellipodia. Nat Rev Mol Cell Biol 2008;9:730–736.

2 Horwitz AR, Parsons JT: Cell migration – movin' on. Science 1999;286:1102–1103.

3 Ishii Y, Yanagida T: A new view concerning an acto-myosin motor. Cell Mol Life Sci 2002;59:1767–1770.

4 Krendel M, Mooseker MS: Myosins: tails (and heads) of functional diversity. Physiology (Bethesda) 2005;20:239–251.

5 Redowicz MJ: Myosins and pathology: genetics and biology. Acta Biochim Pol 2002;49:789–804.

6 Berg JS, Powell BC, Cheney RE: A millennial myosin census. Mol Biol Cell 2001;12:780–794.

7 Dantzig JA, Liu TY, Goldman YE: Functional studies of individual myosin molecules. Ann NY Acad Sci 2006;1080:1–18.

8 Sweeney HL: Regulation and tuning of smooth muscle myosin. Am J Respir Crit Care Med 1998; 158:S95–99.

9 Van Dijk J, Knight AE, Molloy JE, Chaussepied P: Characterization of three regulatory states of the striated muscle thin filament. J Mol Biol 2002;323: 475–489.

10 Yang Z, Sweeney HL: Restoration of phosphorylation-dependent regulation to the skeletal muscle myosin regulatory light chain. J Biol Chem 1995;270: 24646–24649.
11 Hirano K: Current topics in the regulatory mechanism underlying the Ca^{2+} sensitization of the contractile apparatus in vascular smooth muscle. J Pharmacol Sci 2007;104:109–115.
12 Matsumura F: Regulation of myosin II during cytokinesis in higher eukaryotes. Trends Cell Biol 2005; 15:371–377.
13 Entschladen F, Drell TL, Palm D, Bastian P, Pottnoff S, Zaenker KS, Lang K: A comparative review on leukocyte and tumor cell migration with regard to the regulation by serpentine receptor ligands. Signal Transduction 2005;1/2:9–18.
14 Entschladen F, Lang K, Drell TL, Joseph J, Zaenker KS: Neurotransmitters are regulators for the migration of tumor cells and leukocytes. Cancer Immunol Immunother 2002;51:467–482.
15 Neves SR, Ram PT, Iyengar R: G protein pathways. Science 2002;296:1636–1639.
16 Nichols DE, Nichols CD: Serotonin receptors. Chem Rev 2008;108:1614–1641.
17 Pierce KL, Premont RT, Lefkowitz RJ: Seven-transmembrane receptors. Nat Rev Mol Cell Biol 2002;3:639–650.
18 Luttrell LM, Ferguson SS, Daaka Y, Miller WE, Maudsley S, Della Rocca GJ, Lin F, Kawakatsu H, Owada K, Luttrell DK, Caron MG, Lefkowitz RJ: β-Arrestin-dependent formation of β_2-adrenergic receptor-Src protein kinase complexes. Science 1999; 283:655–661.
19 Panchenko MP, Saxena K, Li Y, Charnecki S, Sternweis PM, Smith TF, Gilman AG, Kozasa T, Neer EJ: Sites important for $PLC\beta_2$ activation by the G protein βγ subunit map to the sides of the β propeller structure. J Biol Chem 1998;273:28298–28304.
20 Rebecchi MJ, Pentyala SN: Structure, function, and control of phosphoinositide-specific phospholipase C. Physiol Rev 2000;80:1291–1335.
21 Hofmann J: The potential for isoenzyme-selective modulation of protein kinase C. FASEB J 1997;11: 649–669.
22 Dutta D: Mechanism of store-operated calcium entry. J Biosci 2000;25:397–404.
23 Takai Y, Sasaki T, Tanaka K, Nakanishi H: Rho as a regulator of the cytoskeleton. Trends Biochem Sci 1995;20:227–231.
24 Lorenzo PS, Beheshti M, Pettit GR, Stone JC, Blumberg PM: The guanine nucleotide exchange factor RasGRP is a high-affinity target for diacylglycerol and phorbol esters. Mol Pharmacol 2000;57: 840–846.
25 Griffioen G, Thevelein JM: Molecular mechanisms controlling the localisation of protein kinase A. Curr Genet 2002;41:199–207.
26 Ellerbroek SM, Wennerberg K, Burridge K: Serine phosphorylation negatively regulates RhoA in vivo. J Biol Chem 2003;278:19023–19031.
27 Laudanna C, Campbell JJ, Butcher EC: Elevation of intracellular cAMP inhibits RhoA activation and integrin-dependent leukocyte adhesion induced by chemoattractants. J Biol Chem 1997;272:24141–24144.
28 Pollard TD, Korn ED: *Acanthamoeba* myosin. I. Isolation from *Acanthamoeba castellanii* of an enzyme similar to muscle myosin. J Biol Chem 1973;248:4682–4690.
29 Mermall V, Post PL, Mooseker MS: Unconventional myosins in cell movement, membrane traffic, and signal transduction. Science 1998;279:527–533.
30 De La Cruz EM, Ostap EM: Relating biochemistry and function in the myosin superfamily. Curr Opin Cell Biol 2004;16:61–67.
31 Falk DL, Wessels D, Jenkins L, Pham T, Kuhl S, Titus MA, Soll DR: Shared, unique and redundant functions of three members of the class I myosins (MyoA, MyoB and MyoF) in motility and chemotaxis in *Dictyostelium*. J Cell Sci 2003;116:3985–3999.
32 Novak KD, Titus MA: Myosin I overexpression impairs cell migration. J Cell Biol 1997;136:633–647.
33 Stafford WF, Walker ML, Trinick JA, Coluccio LM: Mammalian class I myosin, Myo1b, is monomeric and cross-links actin filaments as determined by hydrodynamic studies and electron microscopy. Biophys J 2005;88:384–391.
34 Golomb E, Ma X, Jana SS, Preston YA, Kawamoto S, Shoham NG, Goldin E, Conti MA, Sellers JR, Adelstein RS: Identification and characterization of non-muscle myosin II-C, a new member of the myosin II family. J Biol Chem 2004;279:2800–2808.
35 Katsuragawa Y, Yanagisawa M, Inoue A, Masaki T: Two distinct non-muscle myosin-heavy-chain mRNAs are differentially expressed in various chicken tissues. Identification of a novel gene family of vertebrate non-sarcomeric myosin heavy chains. Eur J Biochem 1989;184:611–616.
36 Horowitz A, Trybus KM, Bowman DS, Fay FS: Antibodies probe for folded monomeric myosin in relaxed and contracted smooth muscle. J Cell Biol 1994;126:1195–1200.
37 Wang F, Kovacs M, Hu A, Limouze J, Harvey EV, Sellers JR: Kinetic mechanism of non-muscle myosin IIB: functional adaptations for tension generation and maintenance. J Biol Chem 2003;278: 27439–27448.

38 Bastian P, Lang K, Niggemann B, Zaenker KS, Entschladen F: Myosin regulation in the migration of tumor cells and leukocytes within a three-dimensional collagen matrix. Cell Mol Life Sci 2005;62:65–76.
39 Eddy RJ, Pierini LM, Matsumura F, Maxfield FR: Ca^{2+}-dependent myosin II activation is required for uropod retraction during neutrophil migration. J Cell Sci 2000;113:1287–1298.
40 Lammermann T, Bader BL, Monkley SJ, Worbs T, Wedlich-Soldner R, Hirsch K, Keller M, Forster R, Critchley DR, Fassler R, Sixt M: Rapid leukocyte migration by integrin-independent flowing and squeezing. Nature 2008;453:51–55.
41 Nakayama M, Amano M, Katsumi A, Kaneko T, Kawabata S, Takefuji M, Kaibuchi K: Rho-kinase and myosin II activities are required for cell type and environment-specific migration. Genes Cells 2005;10:107–117.
42 Entschladen F, Drell TL, Lang K, Masur K, Palm D, Bastian P, Niggemann B, Zaenker KS: Analysis methods of human cell migration. Exp Cell Res 2005;307:418–426.
43 Trybus KM: Myosin V from head to tail. Cell Mol Life Sci 2008;65:1378–1389.
44 Veigel C, Wang F, Bartoo ML, Sellers JR, Molloy JE: The gated gait of the processive molecular motor, myosin V. Nat Cell Biol 2002;4:59–65.
45 Espindola FS, Banzi SR, Calabria LK, Custodio RJ, Oliveira RA, Procopio LD, Lima AB, Cunha-Junior JP, Coelho MV, Guedes IM, Pellizzon CH, Larson RE, Espreafico EM: Localization of myosin Va in subpopulations of cells in rat endocrine organs. Cell Tissue Res 2008;333:263–279.
46 Al-Haddad A, Shonn MA, Redlich B, Blocker A, Burkhardt JK, Yu H, Hammer JA, 3rd, Weiss DG, Steffen W, Griffiths G, Kuznetsov SA: Myosin Va bound to phagosomes binds to F-actin and delays microtubule-dependent motility. Mol Biol Cell 2001; 12:2742–2755.
47 Provance DW, Mercer JA: Myosin-V: head to tail. Cell Mol Life Sci 1999;56:233–242.
48 Lan L, Han H, Zuo H, Chen Z, Du Y, Zhao W, Gu J, Zhang Z: Upregulation of myosin Va by snail is involved in cancer cell migration and metastasis. Int J Cancer 2009 (in press).
49 Buss F, Spudich G, Kendrick-Jones J: Myosin VI: cellular functions and motor properties. Annu Rev Cell Dev Biol 2004;20:649–676.
50 Wells AL, Lin AW, Chen LQ, Safer D, Cain SM, Hasson T, Carragher BO, Milligan RA, Sweeney HL: Myosin VI is an actin-based motor that moves backwards. Nature 1999;401:505–508.
51 Geisbrecht ER, Montell DJ: Myosin VI is required for E-cadherin-mediated border cell migration. Nat Cell Biol 2002;4:616–620.
52 Buss F, Luzio JP, Kendrick-Jones J: Myosin VI, an actin motor for membrane traffic and cell migration. Traffic 2002;3:851–858.
53 Yoshida H, Cheng W, Hung J, Montell D, Geisbrecht E, Rosen D, Liu J, Naora H: Lessons from border cell migration in the *Drosophila* ovary: a role for myosin VI in dissemination of human ovarian cancer. Proc Natl Acad Sci USA 2004;101:8144–8149.
54 Dunn TA, Chen S, Faith DA, Hicks JL, Platz EA, Chen Y, Ewing CM, Sauvageot J, Isaacs WB, De Marzo AM, Luo J: A novel role of myosin VI in human prostate cancer. Am J Pathol 2006;169:1843–1854.
55 Tuxworth RI, Weber I, Wessels D, Addicks GC, Soll DR, Gerisch G, Titus MA: A role for myosin VII in dynamic cell adhesion. Curr Biol 2001;11:318–329.
56 Tuxworth RI, Stephens S, Ryan ZC, Titus MA: Identification of a myosin VII-talin complex. J Biol Chem 2005;280:26557–26564.
57 Redowicz MJ: Unconventional myosins in muscle. Eur J Cell Biol 2007;86:549–558.
58 Bennett RD, Mauer AS, Strehler EE: Calmodulin-like protein increases filopodia-dependent cell motility via upregulation of myosin-10. J Biol Chem 2007;282:3205–3212.
59 Zhang H, Berg JS, Li Z, Wang Y, Lang P, Sousa AD, Bhaskar A, Cheney RE, Stromblad S: Myosin-X provides a motor-based link between integrins and the cytoskeleton. Nat Cell Biol 2004;6:523–531.
60 Yonezawa S, Yoshizaki N, Sano M, Hanai A, Masaki S, Takizawa T, Kageyama T, Moriyama A: Possible involvement of myosin-X in intercellular adhesion: importance of serial pleckstrin homology regions for intracellular localization. Dev Growth Differ 2003;45:175–185.

Prof. Dr. Frank Entschladen
Institute of Immunology, Witten/Herdecke University
Stockumer Strasse 10, DE–58448 Witten (Germany)
Tel. +49 2302 926 187, Fax +49 2302 926 158, E-Mail entschladen@uni-wh.de

Author Index

Subject Index